Ωηcε υpο
a ηumβεɾ

Ωηce upoη a ηumβer

The Hidden Mathematical Logic of Stories

John Allen Paulos

BASIC
BOOKS

A Member of the Perseus Books Group

Copyright © 1998 by John Allen Paulos.
Published by Basic Books,
A Member of the Perseus Books Group

FIRST EDITION

Designed by Heather Hutchison

Library of Congress Cataloging-in-Publication Data
Paulos, John Allen.
    Once upon a number : the hidden mathematical logic of stories /
John Allen Paulos.
        p.   cm.
    Includes index.
    ISBN 0-465-05158-8 (hc); ISBN 0-465-05159-6 (pbk.)
    1. Mathematical statistics. 2. Logic, Symbolic and mathematical.
I. Title.
QA276.P285   1998
519.5—dc21   98-39252
                                                                  CIP

99 00 01 02 ❖/RRD 10 9 8 7 6 5 4 3 2 1

*To that most astute philosopher, David Hume,
who wrote, "I cannot but consider myself as a kind of
resident or ambassador from the dominions of learning to
those of conversation, and shall think it my constant
duty to promote a good correspondence betwixt these two states,
which have so great a dependence on each other."*

coŋteŋts

Once upon
a number

introduction

It would be exaggerating to say that our relationship is hostile; I live, I let myself live, so that Borges can weave his literature and that literature justifies me. . . . I don't know which of us is writing this page.

—Jorge Luis Borges

AT HIS FAINT CHUCKLE she turned and faced her once-beloved uncle. Unceremoniously she ripped the papers from the pocket of his Hawaiian shirt as he nervously backed away toward the hotel room door, and with unmitigated disgust at both his blubber and his duplicity she hissed, "Twenty-two point eight percent of all bankruptcies filed between July 1995 and June 1997 were attributed to bad legal advice, up nine point two percent over the last biennial period."

"I did the best I could," the 273-pound man answered faintly. He was desperate to avoid further rousing his enraged niece, who despite her lithe figure, 113 pounds, and angelic face was capable of inflicting severe damage. Once safely in the hallway, however, he took heart and offered, "A meta-analysis of several studies suggests that fewer than forty percent of legal malpractice cases are due to malicious intent, the balance to simple incompetence." At this she lunged at him, tearing into his thick neck with strong, sharp fingers and ripping the shirt from his bloodied back.

As this vignette shows, stories we tell in everyday life often coexist uncomfortably with statistics of supposed relevance, even when the two do not ostensibly contradict one another. Our stories are filled with people who do things out of desire, fear—and possibly an unnatural love of rigatoni. Each particular circumstance and situation looms large in every description. In statistics, however, there are rarely agents; only demographics, general laws, processes. Particularities and details are usually dismissed as unimportant.

The disjunction between narratives and numbers ranges from the commonplace—mistaking a correlation for a causal connection—to the abstruse. One recent, unusual instance of the collision between our desire for comprehensible stories and a simultaneous attraction to impersonal statistics is provided by the Bible codes phenomenon. The craze began when Eliyahu Rips and two other Israeli mathematicians published a paper in a statistical journal that seemed to suggest that the Torah—the first five books of the Bible—contained many so-called equidistant letter sequences, or ELS, that pointed to significant relations among people, events, and dates.

An ELS is a sequence of letters (Hebrew in this case) each separated from the previous by a fixed interval of other letters. The words of the text are run together and the spaces between them ignored. Thus the English word *generalization* contains an ELS for "Nazi"—*geNerAliZatIon*—in which the fixed interval is only of length 3. (Commonly, ELS intervals are much longer—say the

23rd, 46th, 69th, 92nd letter, and so on, after some initial letter.) The paper found that ELSs of (some variants of) the names of famous rabbis who lived centuries after biblical times and that of their birthdays were often close together in the Torah text, and that the probability of this was minuscule.

The publication of this paper was viewed by the journal's editors as a sort of mathematical puzzle: what, among the many things that might account for this low probability, actually does? This, however, was not how the paper was received. Various groups pounced on this "evidence" as they had on previous Christian and Islamic numerological findings, and pronounced it proof of divine inspiration for the Torah. Michael Drosnin's international best-seller *The Bible Code* went even further and claimed to find in the Torah a prophecy of Itzhak Rabin's assassination and other contemporary events. Not surprisingly, also present is the perennial Kennedy connection, an ELS for *Kennedy* not being far from one for *Dallas*. Although I will discuss the resolution of this and the simple mathematics underlying Bible codes later in the book, the point here is that our hunger for stories, agents, and motives is so strong that contextless sequences of letters are seen by many as teeming with significance.

The abovementioned snippets illustrate only two bad ways in which stories and statistics are bridged. This book is about more intelligent ways of spanning and exploring this gap. The nontechnical question of how we fit both stories and statistics into our lives also is discussed; how we answer it helps define who we are.

Nearly everyone has seen urbanocentric posters of New York (or some other city) with the region's attractions in the foreground and the rest of the world vanishing to a point in the distant background. Our psychological worlds are similarly egocentric: other people form the background for our lives—and most annoyingly, we form theirs. How can all these parochial posters and self-conceptions be reconciled with accurate maps, external complexities, and the disembodied view from nowhere?

Again the question is, to what extent can the logical and psychological gap between stories and statistics—and the related gaps between subjective viewpoint and impersonal probability, informal discourse and logic, and meaning and information—be closed, or at least clarified? There exists a similar uneasy complementarity between literature and science. Literary discussions of individual perspectives, possible scenarios, useful archetypes, and singular oddities are awkwardly paired with scientific talk of objectivity, definitive outcomes, universal truths, and general cases. Along slightly different lines, terms such as *lady luck* and *miracle* are uncomfortably arrayed against *chance* and *coincidence.*

Can we successfully straddle the chasm between people who see the world exclusively in terms of good guys and bad guys and those who see it in terms of chance and number, between "literary" and "scientific" culture, and, at two extremes, between conspiracy theorists and "nowhere men"? In an increasingly webified world, can purely personal perspectives and attitudes that do not undervalue scientific objectivity gain respectable status? If so, how do we square the fact that almost everyone feels personally aggrieved yet almost no one deems himself an aggriever?

How, for instance, do we draw a coherent picture that encompasses both human meanings and fragmented bits of information? In what ways do stories (say, of the woman and her corpulent uncle) and statistics (say, of legal malpractice) prove cogent and binding? Are not statistical notions refinements and distillations of ideas suggested by repetitive stories and events? What are the narrative implications of the mathematical notions of complexity and "order for free"? What do interpretations of literature have in common with applications of statistics? What does literary criticism have to do with cryptography?

In the following connected essays I hope to cast an oblique but penetrating light on these questions. Some of the oddities and problems associated with the obvious gulf between stories and statistics (such as mistaking anecdotes for statistical evidence or, conversely, taking averages to be descriptive of individual cases) are the result of applying a logic appropriate in one domain to another, quite disparate, one. Unlike the logic of mathematics and the physical sciences, the truths of informal, everyday logic depend critically on context and on the individual, nonsubstitutable aspects of any situation. Projecting specific conditions of a game or activity, say, or particular religious beliefs onto the physical universe—or, conversely, deriving strategies for such a game or activity or for an individual's religious beliefs from the laws of physics—is only one example of this confusion of domains. So too, in a slightly different way, is the ascribing of significance to biblical codes.

The relationship between personal and objective is often subtle. How we choose to define problems or issues affects their resolution; this occurs, for example, when personally chosen lottery numbers turn out to be more likely winners than machine-picked numbers (even though any set of numbers is as likely to be chosen by the state as any other). More generally, the way in which we are inextricably linked by our common knowledge and implicit understandings points to some interesting extensions of standard mathematical practice.

In between vignettes and parables I will discuss alternative logics, ideas from probability and statistics, codes and information theory, the philosophy of science, and a smidgen of literary theory and use them to limn the intricate connections between two fundamental ways of relating to our world—narratives and numbers. Bridging this gap has been, in one way or another, a concern in all my previous books. It is, I think, a concern of 63.21 percent of us.

1

Between stories and statistics

> "Now a bismuth isotope is going to come out!" I said
> hastily, watching the newborn elements crackle forth from
> the crucible of a "supernova" star. "Let's bet!"
>
> **—Italo Calvino**

Stories and statistics? Whatever might this juxtaposition be getting at? Literature by the numbers? Features in the sporting news? Biographies of Harris, Field, Gallup, and Yankelovich? If pressed, most people would probably say something dismissive, such as stories and statistics go together like a horse and paper clip, or to preserve the alliteration, like stamps and stogies; yet this book takes the relationship seriously.

One of its presuppositions is that storytelling and informal discourse have given birth over time to the complementary modes of thinking employed in statistics, logic, and mathematics generally. Although the latter skills are perhaps more difficult to come by and may even run counter to our intuitions, we can say that first, we tell stories, and then—in the blink of an eon—we cite statistics.

There are a number of vaguely similar "obstetric" relationships: particular versus general, subjective versus universal, intuition versus proof, drama versus the timeless, first person versus third person, special versus standard. The first element in each pairing, while it may be held in lower regard, gives rise to or provides the ground for the second. Thus, a feeling of subjectivity is a necessary preliminary for an appreciation of universality, and dramatic immersion in the moment gradually leads to an awareness of the timeless.

Thinking of these oppositions in a naturalistic way suggests that the chasm between them is more a matter of tradition, degree, and terminology than something untraversably deep. I believe this to be so; and because the gap between stories and statistics is a synecdoche for the better-known gap between what C. P. Snow has deemed two cultures—the literary and the scientific—some of my points may have broader resonance than initial appearances would indicate. (I will sometimes use *statistics* in a quite extended sense.) Since *synecdoche* is a literary term for a figure of speech in which the part is substituted for the whole, or sometimes the other way around, its use is somewhat analogous to substituting a sample for the whole population. With this bit of pedantry we have already landed our first piece of kite string across the chasm.

PRIMITIVE GLIMMERINGS

Notions of probability and statistics did not suddenly appear in the full dress regalia we encounter in mathematics classes. There were plebeian glimmerings of the concepts of average and variability in stories dating from antiquity. Bones and rocks were already in use as dice. References to likelihood appear in early literature. For some at least, the importance of chance in everyday life was clearly understood. It is not hard to imagine thoughts of probability flitting through our ancestors' minds. (If I'm lucky, I'll get back before they finish eating the beast; it doesn't seem likely that they would leave his cattle untouched, but steal his collection of acorns; he usually exaggerates his kill.)

Millennia later ideas of chance and probability were formalized as Pascal and Fermat refined them to solve certain gambling problems in the seventeenth century. Laplace and Gauss further developed their applications to scientific concerns in the next century and a half; and Quetelet and Durkheim used them in the nineteenth century to help understand regularities in social phenomena. (The chances of getting at least one 6 in four rolls of a single die are greater than the chances of getting at least one 12 in twenty-four rolls of a pair of dice; the probability of a particle decaying in the next minute is .927; exit polls show that 4 out of 5 voters in favor of gun control legislation cast their ballots for Gore.)

After this bullet train through the history of statistics, let me slow down to note some of the many colloquial ancestors of the most salient ideas in probability and statistics. Consider first the notions of central tendency: average, median, mode, et cetera. They most certainly grew out of such workaday words as *usual, customary, typical, same, middling, most, standard, stereotypical, expected, non-descript, normal, ordinary, medium, conventional, com-*

monplace, so-so, and so on. It is hard to imagine prehistoric humans—even those lacking the vocabulary above—not possessing some rudimentary idea of the typical. Any situation or entities such as storms, animals, or rocks that occurred again and again would, it seems, lead naturally to the notion of a typical or average recurrence.

Or examine the precursors of the notions of statistical variation: standard deviation, variance, and the like. These are words such as *unusual, peculiar, strange, singular, original, extreme, special, unlike, unique, deviant, dissimilar, disparate, different, bizarre, too much,* and so on. The slang term *far-out* to indicate unconventionality is particularly interesting, because an observation that is far out on the "tail" of a graph of a statistical distribution is rare and unusual, and bespeaks a high degree of variability in the quantity in question. Over time, any recurrent situation or entity would suggest the notion of an unusual exception. If some events are common, others are rare.

Probability itself is present in such words as *chance, likelihood, fate, odds, gods, fortune, luck, happenstance, random,* and many others. Note that mere acceptance of the idea of alternative possibilities and open-endedness essential to storytelling almost entails some notion of probability; some scenarios will be judged more likely than others. The need to single out aspects of recurring situations and entities leads to the key statistical concept of sampling as well, reflected in words such as *instance, case, example, cross-section, observation, specimen,* and *swatch.* Likewise, the natural mental process of yoking together like things suggests the important idea of correlation, which has the following correlates (so to speak): *association, connection, relation, linkage, conjunction, conformity, dependence, proportionate,* and the ever too ready *cause.*

As R. P. Cuzzort and James Vrettos demonstrated in *The Elementary Forms of Statistical Reason,* even less familiar statistical ideas such as control, standardization, hypothesis testing, so-called Bayesian analysis (how we revise our probability estimates in light of new evidence), and categorization correspond to commonsense phrases and ideas that are an integral part of human cognition and storytelling. Like Molière's character who is shocked to find he has been speaking prose his whole life, many people are surprised when told that much of what they characterize as common sense is statistics, or more generally, mathematics. It is also telling that the word *account* refers not only to numbers but to narratives as well.

Admit it or not, we are all statisticians, as when we make grand inferences about a person from that tiny sample of behavior known as a first impression. The difference between mathematical statistics and the everyday variety often is simply the degree of formalization and objective rigor. Standard deviation is computed according to specific rules and definitions, as are correlation coefficients, the rank-sum statistic, chi-square values, and averages (what these are is not here important, although I maintain that they can be communicated via stories and common situations); their everyday cousins are not so formalized.

There may be constraints on the mundane use of these terms as well. The comedian Steven Wright tells a story about going into a clothing store and telling the clerk he's looking for a shirt that is "extra medium." I've expropriated this remark a number of times (usually at ice-cream parlors) and have found that it usually elicits temporary confusion—evidence that people appreciate that the formal properties of an average make the phrase extra redundant. (Or perhaps the response is to my being extra annoying.) Likewise, people recognize the humor of Garrison Keillor's Wobegon effect, according to which almost everybody is above average, or that of a recent headline in a West Virginia newspaper that read "Area Jobless Rate Up, But Still at Record Low." Empty commentary such as "Surveys show that among some voters there is support for the initiative," which I heard recently on a local radio station, provides another example: except for initiatives that are unanimously hated, this is always the case.

The great French mathematician Laplace wrote, "The theory of probabilities is at bottom nothing but common sense reduced to calculus." Voltaire, his much older contemporary, added, "Common sense is not so common."

Stories as Context for Statistics

Unfortunately, people generally ignore the connections between the formal notions of statistics and the informal understandings and stories from which they grow. They consider numbers as coming from a different realm than narratives and not as distillations, complements, or summaries of them. People often cite statistics in bald form, without the supporting story and context* needed to give them meaning.

---

*Not only is context a link between stories and statistics, but also the word *context* itself, if we force things a bit by conjoining *conte*, a short tale or adventure, with *xt*, the most common variables used in statistics and mathematics generally, can be seen as bridging the two realms.

Part of context is internal and attitudinal. As will be discussed in a later chapter, people don't fully realize that how *we* characterize people and events, how we view their circumstances and context, and how we imbed them into stories often determines to a large extent what we think of them. For example, if we describe a person, Waldo, as coming from country X, 45 percent of whose citizens have a certain characteristic, then it seems reasonable to assume (if we know nothing else about him) there is a 45 percent probability that Waldo shares this characteristic. But if we describe Waldo as belonging to a certain ethnic group, 80 percent of whose members in the region comprising countries X, Y, and Z have the characteristic in question, then we will most likely conclude the chances are 80 percent that Waldo shares this characteristic. And if we describe Waldo as belonging to a nation-X-wide organization, only 15 percent of whose members have this characteristic, then we are likely to state that his chances of having the characteristic are only 15 percent. Which (combination of) descriptions we employ to an extent is up to us, so the pleasingly precise statistics we confidently cite are as revealing of us as they are of Waldo (who, just for the record, does not have the characteristic).

More commonly, the problem is not with our attitudes but with our knowledge. We are simply unaware of the external context of most statistics we read or hear about. The contextual questions we should ask when reading news stories, for example, are the very ones statisticians ask when presented with a survey of some sort. We want answers to questions such as how many, how likely, and what percentage, of course. But we also want to know whether the numbers on homelessness or child abuse, say, come from police blotter reports (in which case they are likely to be low) or whether they come from scientifically controlled studies (in which case they are likely to be somewhat higher) or whether they come from the press releases of groups with an ideological axe* to grind (in which case they are liable to be extremely high—or extremely low, depending on the ideology).

Without an ambient story, background knowledge, and some indication of the provenance of the statistics, it is impossible to evaluate their validity. Common sense and informal logic are as essential to this task as an understanding of the formal statistical notions; both are preconditions for numeracy. Although many stories need no numbers, some accounts without supporting statistics run the risk of being dismissed as anecdotal. Conversely, while some figures are almost self-explanatory, statistics without any context always run the risk of being arid, irrelevant, even meaningless.

Consider two recent items in the news, the Consumer Price Index and the birth order effect among siblings.

---

*A recent study by a group opposed to the death penalty stated that blacks convicted of murder in Philadelphia faced four times the odds of being sentenced to death as did others so convicted. Although racial sentencing disparities are certainly troubling, using the odds of an event (technically defined as the ratio of the probability that it occurred divided by the probability that it didn't) greatly overstates them. For example, if 99% of blacks convicted of an especially brutal sort of murder were sentenced to death and 96% of others so convicted were, the odds faced by blacks (.99/.01) would be four times those faced by others (.96/.04).

Understanding the CPI's considerable impact on the economy requires that one have an appreciation not only of rates and exponential growth but also economic theory, taxation codes, partisan politics, and psychology. Many economists have suggested that the CPI, which tracks the price of a relatively fixed basket of consumer goods, gives too high an estimate for inflation and will cost the government hundreds of billions of dollars over the next decade in increased costs of programs and reduced tax revenues. Surprisingly, the argument does not involve mathematics or even economics so much as it does psychology. Many believe the overestimate results from the fact that the CPI ignores improvements in the quality of goods (televisions and cars, for example), the introduction of new goods (the notebook computer on which I'm writing), and the substitution of one good in the basket for another not in the basket (chicken for beef when the latter's price rises). The CPI's alleged overestimation is a *story* in which mathematics plays an important role, but one in which the issues of tax law, social practice, and personal psychology provide the essential context.

A similar point can be made about the birth order effect, the topic of a book by Frank Sulloway in which he maintains that despite sharing 50 percent of their DNA siblings differ systematically from each other owing to their order of birth. Sulloway ascribes this difference to family dynamics: firstborns establish a niche in the family, and to protect it they remain more attentive to their parents' desires and therefore tend to be conservative and supportive of the status quo. Laterborns must find more creative ways to compete with their older siblings for parental favors, and thus tend to be more innovative. Both the topic and the book are huge and statistics plays a key role in Sulloway's argument, but as in the case of the Consumer Price Index, the ambient story and its assumptions are necessary and open to criticism (even if the formal mathematics is unexceptionable).

Why, for example, are only children considered first-borns? They're also the "babies" of their families. Is functional birth order (due to adoption, sibling death, desertion, etc.) a reasonable substitute for biological birth order? How does one decide whether a scientist or political figure (those whom Sulloway studied) should be classified as a conservative or a liberal? What effects might be a consequence of limiting the study to only those historical figures famous enough to have been written about?

Without going into such complex issues, I do want to stress that denial of the mutual dependence of stories and statistics—and the pedagogy that results from such denial—is one reason for the disesteem in which statistics, and mathematics and science generally, are widely held. Its practitioners are simultaneously hailed as awe-inspiring geniuses and summarily dismissed as ivory-tower eccentrics. (Most of the time they are neither, sometimes one or the other, rarely both.) Describing the world may be thought of as an Olympic contest between simplifiers—scientists in general, statisticians in particular—and complicators—humanists in general, storytellers in particular. It is a contest both should win.

SKETCH FOR A
MATHEMATICAL SHORT STORY

Stories not only provide context for statistical statements but can illustrate and vivify them as well:*

A bookish, somewhat nerdy man is telling his kids the Leo Rosten story about the famous rabbi who was asked by an admiring student how it was that the rabbi always had a perfect parable for any subject. The rabbi replied with a parable about a recruiter in the Tsar's army who was riding through a small town and noticed dozens of chalked circular targets on the side of a barn, each with a bullet hole through the bullseye. The recruiter was impressed and asked a neighbor who this perfect shooter might be. The neighbor responded, "Oh that's Shepsel, the shoemaker's son. He's a little peculiar." The enthusiastic recruiter was undeterred until the neighbor added, "You see, first Shepsel shoots and *then* he draws the chalk circles around the bullet hole." The rabbi grinned. "That's the way it is with me. I don't look for a parable to fit the subject. I introduce only subjects for which I have parables."

A stricken look crosses the man's face as he closes the book, hurries his kids off to bed, distractedly bids his wife good night and retreats to his study, where he starts scribbling, making calls, and performing calculations. The idea

---

*As with philosophical abstractions, many of the ideas and problems in probability theory have standard vignettes associated with them. Examples are such stories as the gambler's fallacy and gambler's ruin, the Banach match box problem, the drunkard's random walks, the Monty Hall problem, the St. Petersburg paradox, the random chord problem, the hot hand, the Buffon needle problem, and many others.

for a lucrative con game grows clearer in his mind. The next day he does some research, stops by the post office, and for the next two evenings sends letters to thousands of known sports bettors "predicting" the outcome of a certain sporting event. To half of these people he predicts that the home team will win, to the other half that it will lose. His con depends on the simple fact that whatever happens in the sporting event, he is right for half the bettors.

His wife wonders at their huge postage bills and secret telephone calls and nags him about their worsening financial and marital situation. The following week he sends out more letters and makes another prediction, but this time to only half of the people for whom he has been right; the other half he ignores. To half of this smaller group he predicts a win in another sporting event, to the other half a loss. Again, for half of this group his prediction will be correct, and thus for one-fourth of the original group he will be correct two times in a row. To half of this one-fourth he predicts a win the week after that, to the other half a loss; again he ignores those to whom he has made an incorrect prediction. Once again he is correct, for the third straight time—that is, for one-eighth of the original population. He continues in this way to extend his string of "successful predictions" to a smaller and smaller group of bettors. With great anticipation he then sends a letter to those who are left in which he points out his impressive string of successes and requests a substantial payment to keep these valuable and seemingly oracular "predictions" coming.

He receives many payments and makes a further prediction. Again he is correct for half of the remaining bettors and drops the half for whom he is incorrect. He asks the former for even more money for another prediction, receives it, and continues. Finally, with only a few bettors left, one of them, a rough underworld type, traces the man down, kidnaps him, and demands a prediction on whose strength he plans to bet a lot of money. The kidnapper threatens the man's family, and not understanding how he could be the recipient of so many consecutive correct predictions, refuses to believe that this is a con game. The man makes some interesting philosophical points in an effort to convince the kidnapper that he is not divine. The nerdy scam artist and the muscled extortionist are a study in contrasts: they speak different languages and have different frames of reference but seem to have similar attitudes toward women and money. Under extreme duress the con man makes a prediction that happens to be correct, and the kidnapper, more convinced than ever that he is in control of a money tree, now wants to bet all his assets and those of his associates on the next prediction.

The denouement involves the man's mistress, on account of whom he originally felt the need for the extra cash supplied by his scam. She is instrumental in his escape from his captor before he makes an incorrect prediction that would result in his family being murdered. Using an ingenious code between them they manage to give pause to the kidnapper and scare him from ever bothering them again. In the last scene he is working the same scam, this time with a stock market index since he wants a higher-class clientele. He is married to his mistress but has another on the side, who is beginning to make even bigger money demands. He sits at his desk, doodling little bullseyes and targets on an envelope.

•   •   •

The branching possibilities idea in this sketch comes naturally to a probabilist or statistician, since so-called tree diagrams (which date from Dutch mathematician Christian
Huygens in the late seventeenth century) are useful for determining the probability of sequences of events. But tree
diagrams are also helpful when thinking of the choices fac-
ing characters in stories or when considering plot turns
that are more externally driven. Each path along the
branches of the tree of possibilities (it may help to visualize
the tree growing to the right with time rather than up) corresponds to a sequence of choices by the characters or to
other turns of the plot, while sub-branches and twigs correspond to various digressions and diversions. And so the forward branching, sideward digression, and occasional
backtracking at various levels and scales can be taken as a
model for how we generally tell stories.

This forking of reality suggests the increasingly popular
idea of computer-generated fiction in which linear progression through a story would not be necessary. One
wouldn't read them so much as wander around in them.
Without a conventional story line there would be indefinitely many narrative excursions, not all of them unified
by the consciousness of one protagonist. After reading a
passage, one would proceed forward linearly, backtrack
to a previous passage, or move sideward by focusing
(clicking) on any significant word or phrase in the passage and being directed to an elaboration of it. The
virtue of this arboreal proliferation of digressions presumably would be the evanescent, open-ended, lifelike
feel it would provide the reader–browser.

Ideally, one would read only those developments, asides, and vignettes he or she finds intriguing. It would be interesting if the imagined text/software had a quiz at the end, the answers to which would be dependent on which portions the reader had selected. Even in such a mammoth text as this one would not be able to develop every conceivable fork the tale might take. Artistry is required to overcome the combinatorial explosion of possibilities and seamlessly bind and weave material to create the illusion of free choices and unbounded bifurcation. At crucial junctures, for example, there might be few, if any, alternatives. The effect, like that of a stream of water through a bottleneck, would suggest the protagonist's single-mindedness at such moments.

Done right (and I have not yet seen an example that comes close to being satisfying), the almost sentient matrix of diversion, digression, and horizontal movement within such a work would animate the characters and foster a greater sense of identification with them. Details, large and small, on matters both critical and trivial, would tumble forth from such a multidimensional chronicle and help truly animate its milieu and time. Mathematicians often speculate about what Archimedes, Gauss, Poincare, or other mathematical virtuosos of the past might have accomplished with the checking and searching capacities of a computer. I wonder what Sterne, Joyce, Borges, or others whose works are reminiscent of what I am envisioning might have done with electronic help. For all of its dense sprawl, one might come away from such a text with as vivid and precise a grasp of virtual individuals and their circumstances as it is possible to have.

Of course, such a work might be dismissed as a mere technical curiosity. A more likely obstacle to its imminent creation is the dearth of writers capable of literary nuance and psychological subtlety and also of the architectonic vision and software skills necessary to articulate such a complex branching "story."

THE DIFFERENT SCOPES OF
STORIES AND STATISTICS

Zillions of stories, from the *Iliad* and *Odyssey* to art films and television soap operas, and zillions of surveys, polls, and studies demonstrate the many contrasts between stories and statistics. (*Zillions* is a useful word even for a mathematician; it certainly beats "an indeterminately large number." The words *umpteen* and *oodles* are also useful.) One major difference is that in storytelling the focus is almost always on individuals rather than analyses, arguments, and averages. Such a focus is a necessary corrective to overweening abstraction and keeps the statistics in human perspective.

Even if they are true, to take an extreme example, there is something inhuman and vaguely pornographic about statistics that maintain that since half the people in the United States are men and half are women, the average American adult has one ovary and one testicle. Or that the average resident of Dade County, Florida, is born Hispanic and dies Jewish. Pornography, though, with its loosely bound sequences of storyless sexual couplings (or triplings) often has the feel of a statistical survey.

But a focus on individuals can be deceptive and manipulative and can distort discussions of public policy issues, especially those involving health and safety. A poignant television story of a victim of a rare reaction to a vaccine can render invisible the vast good brought about by this same vaccine. There are countless examples of such media-induced bathos.

Some writers try to enjoy the virtues of both individual accounts and statistical surveys by improperly conflating them. The result doesn't so much bridge the gulf between them as fall into it. A typical example is the convention of conjuring up some "representative" person—a fictional Jeremy, Linda, or Kevin (but never a Waldo or Gertrude)—to endorse or exemplify whatever statistical conclusion a newspaper or magazine article has reached. (Janet Cooke of the *Washington Post* had a Pulitzer Prize taken away from her for taking this practice to extremes.)

A number of other critical aspects of the gap between statistics citing and storytelling derive from the fact that, as the proverbial writing teacher's maxim enjoins, a story shows, rather than tells. Stories may employ dialogue and other devices and do not limit themselves to declarative pronouncements; they develop the context and relevant relationships instead of merely positing raw data; they are open-ended and metaphorical, whereas statistics and mathematics generally are determinate and literal; and stories unfold in time instead of being presented as timeless.

Stories presuppose a particular point of view (or possibly several) rather than offering an agentless impersonal overview as do statistics. Consider, for example, the notion of a probabilistic distribution of the weights of females in a certain population. Via a formula or graph (such as the well-known normal curve, or belly-shaped curve, as one of my students aptly named it), it gives one a god's eye of the fraction of women within any given interval of weights. From the distribution one can read off the heaviest weight, the lightest, the most common, the least common, and much else. All the information is there in one snapshot, but devoid of the draconian diets, ice cream, cookies, gorging, and fasting of any particular woman.

For better and for worse, individual stories are more elemental than statistics and hence more emotionally evocative. Phrases such as "betrayed his wife," "hair blowing in the breeze," and "reeking underarm rot" never appear in scientific studies. Instead we get phrases like "72.6 percent thought," "the correlation between," and "margins of error." Even in a domain as saturated with statistics as baseball, the romance of Babe Ruth's story makes his former records of 60 home runs in a season and 714 in a career somehow grander than the newer records set by Roger Maris and Henry Aaron, respectively (even for one such as I, an erstwhile fan of the erstwhile Milwaukee Braves).

Yet there are amalgams of statistics and stories that do, to an extent, bridge the two. In this fuzzy middle ground we find *Rashomon*-like stories that portray many disparate views of the same set of events. Here too are ensemble stories (like many television series) that interweave accounts of each member of a group of related people, and also San Luis Rey stories that loosely tie together the doings of many unrelated people. The more people or viewpoints considered, however, the flatter and more featureless they must be, and the forward progression of time gradually slows into the cross-sectional nowness of most statistical snapshots and surveys (although there are subdisciplines of statistics—stochastic processes and time series—where the concern is with the evolution of variable quantities through time).

A computer analogy is helpful. If we think of conventional stories as being told from one point of view (just as a serial processor performs one calculation at a time), then statistics may be thought of as providing a view from nowhere in particular (many parallel processors performing simultaneous calculations). Between them are amalgams, which may be thought of as a varying number of variously connected viewpoints (processors). Combining the virtues of these two very different ways of apprehending the world—through stories and statistics—can be considered a literary analogue to a common problem in computer design and architecture.

Too Many Characteristics, Not Enough People

The right balance between depth of characterization and the number of characters, however, isn't always clear. In stories, as in everyday life, we interact with relatively few people personally, but they are real three-dimensional folk (actually, in a mathematical sense they are N-dimensional folk for large values of N). They possess or are associated with an indeterminately large number of possible traits, circumstances, relationships, informal rules, and agreements. We certainly do not know everything about those closest to us (or even about ourselves), yet we are aware implicitly of so many details and richness of context that writing it all down would make us all bad novelists. Even to those we don't know well we can attach a dozen adjectives, a few adverbs, and a couple of anecdotes. Contrast this abundance of personal particulars with most scientific studies where, while there may be a very large number of people (or other data), the people surveyed are flat, having only one or two dimensions—

who they will vote for, whether they smoke, or what brand of soft drink or laxative they prefer.

Stories and statistics offer us the complementary choices of knowing a lot about a few people or knowing a little about many people. The first option leads to the common observation that novels illuminate great truths of the human condition. Novels are multivalent and bursting with ironies, details, and metaphors, while social science and demographic statistics can seem simple-minded and repellingly earnest by comparison. We can easily delude ourselves, however, into thinking that more of a general nature is being revealed to us by a memoir, personal reminiscence, novel, or short story than is truly the case. Biased and small samples are always major problems, of course, but my caveat arises from something more specific: the technical, uneuphonic statistical notion of an adjusted multiple correlation coefficient.

If the number of traits considered is large compared to the number of people being surveyed, there will appear to be more of a relationship among the traits than actually obtains. Imagine a study that examined only two people and two characteristics, say intelligence and shyness. Imagine further a graph with degrees of intelligence on one axis and degrees of shyness on the other, and two points on it corresponding to the two people. If the shyer of the two were more intelligent, there would be a perfect correlation between the two traits and a straight line connecting the two points on the graph. More shy, more intelligent. But if the shyer of the two were less intelligent, there still would be a perfect correlation between the two traits and a straight line pointing in the opposite direction connecting the two points. More shy, less intelligent.

You can find perfect correlations that mean nothing for any three people and three characteristics, and in general for any N people and N characteristics. The number of characteristics need not equal the number of people. Whenever the number of characteristics is a significant fraction of the number of people, the so-called multiple correlation among the characteristics will suggest spurious associations.

To tell us anything useful, multiple correlation analysis must be based on a relatively large number of people and a much smaller number of characteristics. Yet the insights that commonly come from stories and everyday life are precisely the opposite. We each know in a full-bodied way relatively few people, and for these people the number of characteristics, relationships, characteristics of relationships, relationships of characteristics, and so on that we are aware of is indeterminately large. Thus we tend to overestimate our general knowledge of others and are convinced of all sorts of associations (more complicated variants of "more shy, less intelligent") that are simply bogus. By failing to adjust downward our multiple correlation coefficients, so to speak, we convince ourselves that we know all manner of stuff that just isn't so.

Just as stories are sometimes a corrective to the excessive abstraction of statistics, statistics are sometimes a corrective to the misleading richness of stories.

STEREOTYPES, WHIMSY, AND STATISTICAL CONSERVATISM

The alternative in everyday life to probabilistic calculations and explanations is the amorphous "discipline" of common sense and rough appraisal. Rather than presenting rigorous proof or careful calculation for fixed propositions, common sense involves thinking in terms of scenarios and situations, empathizing and identifying with people, responding to conversations and weighing observations, and then finally coming to a tentative, sometimes fickle judgment. The knowledge that results is qualitative, imprecise, and context-bound. Common sense often is couched in the language of probability, but attaching a particular number, a precise probability, to a possible outcome is frequently (81.93 percent of the time) an exercise in fatuity. Yet the specter of unwarranted precision seldom deters those who want to give their hunches an air of scientific respectability.

Rather than invoking precise probabilities, in our everyday approach to life we find it more natural to deal with rules of thumb and approximate categories; in other words, with stereotypes. Although many assume that stereotypes are always evil vestiges of a benighted mind-set, more often they are essential to effective communication and have themselves been unfairly stereotyped (assuming a concept can be treated unfairly). Many stereotypes permit the economy of expression necessary for rapid communication and effective functioning. *Chair* is a stereotype, but one never hears complaints from bar stools, recliners, beanbags, art deco pieces, high back dining room varieties, precious antiques, chaise longues, or kitchen instances of the notion. Stereotypes, of course, admit of all sorts of exceptions that upon further examination in individual cases are easily apparent, but this does not mean they should or even can be universally proscribed. Complexity, subtlety, and precision cost time and money, and these expenditures often are unnecessary and sometimes even obscuring.

Recognition of common stereotypes and knowledge of recurring stereotypical situations such as restaurant behavior, retail purchasing, hygiene practices, audience deportment, and so on are essential for navigating through everyday life. Approaches to artificial intelligence, in particular that of computer scientist Roger Schank and others, have reinforced the observation that we chart our course and communicate with others by invoking common types, scenarios, and scripts as shorthand. Like statistical notions, stereotypes do violence to particular cases and individuals but pay their way by summarizing general information the many exceptions to which would be too time-consuming to note.

Of course, I don't dispute that stereotyping people can stimulate unthinking, cruel, and self-fulfilling prejudice, and I'm most certainly agin that.* And yet when we meet or even catch a glimpse of someone, there is a tendency to construct (all right, *I* tend to construct) an instant biography of that person, and in the process, to make all sorts of immediate appraisals. I still find it hard not to draw far-reaching (and often mistaken) judgments, for example, about a person who uses the phrase "between you and I."

---

*A simple experiment, one among hundreds, illustrates how the difference between the objects in two sets—and presumably people in two groups—can be exaggerated simply by labeling them: Four lines are labeled A, while four slightly longer lines are labeled B. People think the differences in length between the lines in the two sets are bigger than do people presented with the same lines unlabeled. Similar judgments of difference persist when the average length of the lines in the two groups is adjusted to be the same.

But foregoing speculation altogether seems too harsh a solution to the problem of stereotypes. Taking the liberty of reporting a seeming case of prescience on my part, I remember reading the Unabomber's manifesto several years ago on a mountaintop in Maine and guessing from its tone, content, and structure that its author was a mathematician. Later, at the time of his arrest, I wrote an Op-Ed piece for the *New York Times* to that effect. It aroused so much ire among some mathematicians, who thought it besmirched their reputations, that the *Wall Street Journal* devoted a long article to the resulting fracas. In the essay I opined that Theodore Kaczynski's Ph.D. in mathematics* was perhaps not quite as anomalous as it seemed (despite the fact that mathematicians are for the most part humorous sorts, not asocial loners, and that the only time

---

*The Unabomber's manifesto appeared to me to be axiomatic in nature, based on a few fundamental postulates from which all else follows logically. Mathematicians often view themselves as *radical* thinkers (in the literal sense), getting to the root of the matter, and this manifesto, which also showed a mathematician's meticulous attention to detail, had the feel of an extended proof based on a few root presuppositions about what constitutes the good life, e.g., personal control, self-reliance, minimal environmental impact. From these elements it seemed that he attempted to derive a radical alternative theory of society—a non-Euclidean approach, if you will, to our social and psychological problems.

Abstract thinking is another obvious characteristic of mathematics. Although it is a precious commodity too often absent from public debate, it has been associated with various dissociative pathologies, and it is easy to see how one trained in such reasoning and in thrall to an ideal could come to justify vicious and murderous acts as a nebulous "good."

Mathematics is also beautiful, but its aesthetic, minimalist and austere, can blind one to the messiness and contingencies of the real

most of us use the phrase "blow up" is when we consider division by zero).

Even in such rare cases it is difficult to suppress hasty judgments and stereotyping; perhaps it is unwise even to try. Nevertheless, if we attempt to keep judgments tentative and back them up to the extent possible, no great harm is done. Unfortunately, I frequently run into people who make no such effort. They claim with an air of dismissive certainty that someone is a racist, or a secret admirer, or is extraordinarily wealthy, or gay, or something else. Usually these assertions are based on some ineffable complex of traits that simply must be recognized. Unless he or she is well known, the person thought to possess the trait is rarely interviewed or investigated (legitimately) to determine if he or she truly has it; some of the hunches are occasionally discovered to be correct by other means, and this is taken to certify them all.

•   •   •

While stereotypes may be a bridge between statistics and stories, like bridges they are sometimes old, rickety, and unreliable. Statistical conclusions, unlike stereotypes, must undergo stringent tests. This point is usually dis-

---

world. Using mathematical principles to devise a grand socioeconomic theory forces one to simplify; in so doing, one can forget that the mathematical or economic model is not the real world. Reality, like the perfectly ordinary woman in Virginia Woolf's essay "Mr. Bennett and Mrs. Brown," is indefinitely complex and impossible to capture completely in any model.

missed as statistical nit-picking; after all, "everybody knows" whatever is being asserted. I have a version of this bias myself: people who make frequent claims about what everybody knows are fools. But everybody knows this.

Statistical decision-making is a drab, conservative process unlike the spirited snap judgments that characterize personal appraisals. The so-called null hypothesis in statistics is the assumption that the phenomenon, relationship, or hypothesis under observation is not significant but merely the result of chance. To reject the null hypothesis it is conventional to require that the probability of the phenomenon occurring merely by chance be less than 5 percent. (This is the source of the story about the statistician who witnessed the decapitation of twenty-five cows, noted that one survived the ordeal, and dismissed the phenomenon as not significant.) In my peregrinations through this world I've observed that few people regularly make decisions like this in their personal lives; it would be too boring even if such precision were possible. (Having offered a partial defense of stereotypes, I should mention that a common related stereotype of statisticians is that they are people who chose their profession because they couldn't stand the excitement of accounting.)

The idea of boredom suggests yet another difference between stories and statistics. In listening to stories we are inclined to suspend disbelief so as to be entertained, whereas in evaluating statistics we are inclined to suspend belief so as not to be beguiled. In statistics we are said to commit a Type I error when we reject a truth and a Type II error when we accept a falsehood. Of course, there is no way to always avoid both types of error, and we have different error thresholds in different endeavors. Nevertheless, the type of error people feel more comfortable making gives some indication of their intellectual personality type. People who like to be entertained and beguiled and hate the prospect of making a Type I error may be more likely to prefer stories to statistics. Those who do not like being entertained or beguiled and hate the prospect of making a Type II error may be more likely to prefer statistics to stories. In any case, this speculation is a short story with no statistics to back it up, so make of it what you wish.

Although wrong much of the time, we have more confidence in our own gut decisions than we do in public ones. We all (not just right-wing Republicans) distrust decisions that are made far from us. We insist on exacting statistical protocols in public decision making, yet oftentimes accept the sloppiest reasoning from those closest to us. In small groups there is trust and little perceived need for statistics. As Theodore Porter has shown in his *Trust in Numbers*, quantitative methods and controls often arise owing to the political weakness of expert communities and a suspicion of their findings by the larger community. Those anticipating distrust are most likely to undergird their conclusions with substantial statistics, or at least adorn them with fake statistical finery.

The sheer impersonality of statistics is attractive to those who dislike the messiness, intimacy, and (melo)drama of particular stories, situations, and people. Stories, it would seem, appeal more to stereotypical women, statistics to stereotypical men (according to conventional wisdom; I have no statistics on whether this is true of real men and women). The conservatism and impersonality of statistical practices is one source of their trustworthiness, while the whimsy and diversity of personal stories is one source of their appeal.

•　　•　　•

Since probability and statistics are formalizations of our pretheoretical intuitions, they usually accord reasonably well with our gut feelings. Still, these disciplines have developed a life of their own independent of our attitudes and beliefs, and in many situations statistics tells us our gut feelings have been led astray. People best respond in small groups where they can use their folk wisdom about others' intents and purposes and where their psychology gives them insight into others' stereotypical behaviors and actions. It is in this realm too where our narrative intuitions are surest, where a few telling details are often sufficient to sketch in a whole world. I conclude with an example from Leonard Michaels's collection of very, very short stories, *I Would Have Saved Them If I Could.* "The Hand" is a fifty-nine word incompressible psychological nugget almost mathematical in its spareness:

> I smacked my little boy. My anger was powerful. Like justice. Then I discovered no feeling in the hand. I said, "Listen, I want to explain the complexities to you." I spoke with seriousness and care, particularly of fathers. He asked, when I finished, if I wanted him to forgive me. I said yes. He said no. Like trumps.

2

βetweeη subjective viewpoiηt aηd ιmpersoηal probability

He was too tired to appreciate the irony, or coincidence, or whatever it was. There were too many ironies and coincidences. A shrewd person would one day start a religion based on coincidence, if he hasn't already, and make a million.

—Don DeLillo

IF THERE IS A GENETIC DISPOSITION to materialism (in the sense of "matter and motion are the basis of all there is," not "I want more cars and houses"), then I suspect I have it. I remember wrestling with one of my younger brothers when I was about ten or eleven and having an epiphany that the stuff of our two heads wasn't any different in kind from the stuff of the rough rug on which I'd just burned my elbow or the stuff of the chair on which he'd just banged his shoulder. The realization that everything, ultimately, is made out of the same matter, that there is no essential difference between the material composition of me and not-me, was clean, clear, and bracing.

With the solipsism I've since concluded is common among kids, I was sure no one had ever had this piercing intellectual insight. A year or two later I recorded on a piece of paper the following syllogism: Everything is made out of atoms. Atoms can't think. Therefore people can't really think. I folded the paper tightly, covered it with tape, placed it in an empty metal talcum container, sealed the top of the container with plastic glue, and buried it in the backyard next to the swing set for future (unthinking) generations to discover.

This youthful atomism evolved quickly into adolescent atheism, intolerant of just-so tales devoid of evidence. The absence of an answer to the question, What caused, preceded, or created God? made, in my eyes, the existence of the latter Being an unnecessary, antecedent mystery. Why introduce Him? Why postulate a completely nonexplanatory, extra perplexity to explain the already sufficiently perplexing and beautiful world? If one were committed to such unnecessary mysteries, why not introduce even more antecedents, such as the Creator's Creator or even His Great Uncle? The notion of chance (or should I write Chance?) also fascinated me; I recall the effect on me upon hearing the cliché "It's all in the cards" uttered by a friend of my parents to indicate an appreciation of the role of happenstance in life. I had always thought this person rather unintelligent, and the remark, mirroring what I then considered to be my most profound thoughts, shocked me.

As I grew older my preoccupation with a scientific, non-narrative view of things spread from the otherworldly to the mundane. I remember thinking that the frequency of the pops heard in cooking popcorn followed a sort of aural normal distribution—a few early pops, a crescendo, a plateau, and then a decrescendo, succeeded by a few stragglers. In junior high I remember reading that one out of five high school girls was not a virgin, then trying to determine the probability of there being a nonvirgin among a group of high school girls I knew. I began to point out the literal interpretation of street signs (VIOLATORS WILL BE TOWED should be VIOLATORS' CARS WILL BE TOWED; and KEEP LITTER IN ITS PLACE counsels us to leave it on the ground else it lose its status as litter—placed in a garbage can, it's no longer litter, but garbage) and was considered strange for my efforts.

Archimedes maintained that given a fulcrum, a long enough lever, and a place to stand, he could move the Earth. The remark suggests the theoretical nature and transcendental longings of mathematics. Popcorn, nonvirgins, and street signs stood for levers, fulcrums, and the Earth in my humblingly puerile case, but I nevertheless developed a keen appreciation of Archimedes' point.

But that's me. You're no doubt different. Like characters in stories, we each have viewpoints that may appear to be a little off. Even when the topic is mundane rather than metaphysical, there is frequently a large gap between subjective probabilities, the numbers we attach to uncertain events, and more objective assignments of probabilities to these events—especially when they involve us personally. As observers often have noted, people tend to exaggerate the likelihood of events that are new, emotionally stirring, dramatic, or concrete, and tend to underestimate the probability of events that are old, emotionally neutral, boring, or abstract.

Our minority status (as individuals we are the sole member of the smallest possible minority) and idiosyncratic viewpoint also affect our view of personal coincidence. From the geometric and experiential fact of our being at the center of our own stories and at the periphery of others', many of us simultaneously—and somewhat bizarrely—conclude that our lives brim with remarkable events and coincidences and that those of others are rather typical. But we're each unique—just like everyone else.

MINORITY VIEWPOINTS, INDIVIDUALS, AND STATISTICS

Of more general concern than particular networks of personal associations is the way that minority status—a crucial ingredient in many people's identities—colors our view of many societal issues. In an oddly basic way, a minority's viewpoint(s), and even an individual's, may be affected by probability and statistics.

A thought experiment illustrates the point. (The term *thought experiment* denotes an idealized investigation of a phenomenon that tries to capture its essence without becoming bogged down in minutiae.) A supercharged area of American life, race relations, certainly is in need of thought experiments, simplistic though they may be. So let us experiment and assume that contrary to fact, blacks and whites hold positions of equal importance and influence. Assume further that about 10 percent of each group is racist, and that the country is both residentially and professionally integrated. Given these unrealistic assumptions it is not hard to demonstrate that since blacks comprise approximately 13 percent of the population and whites the remaining 87 percent (for these purposes, whites are nonblacks), blacks still would suffer disproportionately from racism.

The chance that a white will run into a black racist in any given encounter with another person is 1.3 percent (10 percent of 13 percent), whereas the likelihood that a black on any given encounter will do so is 8.7 percent (10 percent of 87 percent). This disparity becomes more pronounced as the number of a person's contacts grow.

If a white person encounters 5 people, his or her chances of meeting at least 1 racist are 6.3 percent, while the average number of racists he or she will encounter is .07. By contrast, if a black person encounters 5 people, his or her chances of meeting at least 1 racist are 36.6 percent, while the average number of racists he or she will encounter is .44. If a white person encounters 25 people, his or her chances of meeting at least 1 racist rise to 27.9 percent, while the average number of racists he or she will encounter rises to .33. If a black person encounters 25 people, his or her chances of meeting at least 1 racist rise to 89.7 percent, while the average number of racists he or she will encounter rises to 2.18.

The conclusion is that minority status by itself can make equal opportunity difficult to achieve or maintain. In fact, if the already idealized abovementioned conditions held, but now only 2 percent of whites and 10 percent of blacks were racist, blacks would still encounter more racism than would whites.

•   •   •

Most simple models can be made more realistic by introducing more complicated assumptions. I'd like to add a couple to the example just considered. First, replace racism with bias or general offensiveness. Further, don't assume that individuals are biased or unbiased but rather that they are biased to differing degrees, measured simplistically by percentages. The numbers in this example would have to be massaged a bit, but the same lesson would emerge: members of a minority group would encounter more bias than would members of a larger group. Next, assume that the minority group consists of only a few members, say a family. The effect, of course, would be exacerbated. The family generally would be the recipient of incomparably more bias or overall offensiveness than it displayed to the outside world ("generally" because the family might be quite nasty).

Let us take the scenario one step further and shrink the family to an individual. Again, the individual generally would be the recipient of incomparably more bias or offensiveness than he or she displayed to the outside world (again, "generally" allows for the nasty exception). And we are not only confronted with bias or offensiveness; all of us begin life as small beings and know intimately what it is like to be powerless. Instances of powerlessness are much more common than are instances of bias.

These elementary considerations shed some light on why so many of us feel vulnerable and victimized. While most of us, at least by our own reckoning, try to be kind and considerate to others, we very often find that "they" are thoughtless and rude to us. Part of the explanation derives from arithmetic and probability, but this insight should not be imparted to the snarling driver stepping out of his car to contest your simultaneous discovery of "his" parking spot.

Note that only one of the conventional elements of narrative played a role in the aforementioned analysis: the simple notion of an agent's viewpoint. The richness and complexity of most everyday situations make basic arithmetic insights less visible. Similar observations hold for fictional situations. Seeing events from the point of view of a character within a story or through the eyes of the story's narrator is not conducive to probabilistic reasoning. Almost any ostensibly improbable event is likely to be explained away by the reader as resulting from some unspecified facts or assumptions.

MURPHY'S LAW AND THAT PUT-UPON FEELING

Another instance of the connection between the personalization of events and mild paranoia is Murphy's Law. First formulated by the engineer Edward Murphy, it states that generally whatever can go wrong will go wrong. Contrary to the tongue-in-cheek aspect of this characterization, there is something profound about the phenomenon it describes. In many situations the failure of things to work right is not a result of personal bad luck but a consequence of the complexity and interdependence of many systems.

A homely, counterintuitive example of Murphy's Law comes from probability theory and was recently elaborated on by science writer Robert Matthews. Imagine you have 10 pairs of socks and that despite your best efforts 6 socks disappear. (The solution to the problem of the socks' disappearance and their whereabouts is yet another holy grail.) The question is, what is more likely— that you're lucky and end up with 7 complete pairs (i.e., the 6 missing socks come from only 3 pairs), or that you're unlucky and end up with only 4 complete pairs (i.e., that the 6 missing socks come from 6 different pairs)? The surprising answer is that you're more than 100 times as likely to end up with the worst possible outcome, only 4 pairs (plus 6 single socks), than you are to end up with the best possible outcome, 7 pairs (and no singles). More precisely, the probability of 7 pairs is .003, of 6 pairs .130, of 5 pairs .520, and of only 4 pairs .347.

The solution (details of which I will omit) derives from the notion of statistical independence, which is of fundamental interest and thus worth a digression. Two events are said to be independent when the occurrence of one of them does not make the occurrence of the other more or less probable. If you flip a coin twice, each flip is independent of the other. If you choose two people from the phone book, the birth month of one is independent of that of the other. Calculating the probability of two independent events occurring is easy to do: simply multiply their respective probabilities. Thus, the probability of obtaining two heads is $1/4$: $1/2 \times 1/2$. The probability that two people chosen from the phone book are each born in July is $1/144$: $1/12 \times 1/12$. This multiplication principle for probabilities may be extended to sequences of events (as in the aforementioned segment on racism*). The probability of a die's turning up 3 on four consecutive rolls is $(1/6)^4$; of a coin's landing heads six times in a row, $(1/2)^6$; of someone's surviving three shots in Russian roulette, $(5/6)^3$.

The tendency of socks to shed their mates is certainly Murphy's Law with a vengeance! Nevertheless, we should expect this to happen and need not invoke bad luck to account for our incomplete pairs of socks. I realize that most people who speak of a series of minor personal mishaps are merely embellishing a story or trying to establish rapport with others and don't necessarily subscribe to their pronouncements, just as most who refer to the bogeyman don't believe in one. Still, we often feel genuinely perplexed and that the world is conspiring against us, and mathematical debunkings help dispel this illusion.

---

*Thus the probability that a white person will *not* meet a racist in any given encounter is equal to .987, or $(1 - .013)$, since .013 is the probability he or she will meet a racist in any given encounter with another person. And so the probability of not meeting a racist in 5 independent encounters with different people is $(.987)^5$. Since $(.987)^5$ is equal to .937, the probability that a white person will meet a racist in 5 encounters is approximately equal to $(1 - .937)$, or .063. The other cases are calculated similarly.

Self-aggrandizement is the key to the appeal of the illusion and of paranoia in general. The feeling is a consequence of the perhaps subconscious inference that if the world is out to get me, I must be pretty important. It's hard for these people to come to grips with the likely fact that almost no one gives a pair of socks about them.

In an increasingly interconnected world of growing complexity, sometimes it takes relatively little to bring a system down. A pair of socks is "brought down" if a sock is lost. A large collection of parts connected more or less in series (so that if one fails, the system does) is even more vulnerable, as are our physical bodies. When such failures occur (and I would argue that they include diseases and accidents as well as unmatched socks) the stories we tell ourselves and believe genuinely matter. Murphy's Law aptly illustrates an aspect of this nexus among stories, selves, and statistics.

●   ●   ●

Another instance of Murphy's Law is the waiting time paradox. Let's assume you are marooned in some small outpost in the Sahara and are told that on average there are two buses per day that will take you back to civilization. If you arrive at the post at some random time, and if the buses come every 12 hours, then your average waiting time, were you to be regularly marooned there, would be 6 hours (sometimes under 1 hour if you're lucky, sometimes over 11 hours if you're not, sometimes 4 hours, sometimes 8 hours, etc.), which would be a reasonable estimate for your waiting time on this occasion as well.

But if the bus arrival times vary your average wait will be longer. Say, for example, that one of the two buses always arrives at midnight, the other at 2 A.M., and that again you show up at the post at some random time. If you show up in the two-hour interval between midnight and 2 A.M., which will happen about one-twelfth of the time, your average wait will be 1 hour. If you show up in the twenty-two-hour interval between 2 A.M. and midnight, which will happen about eleven-twelfths of the time, your average wait will be 11 hours. Putting this all together yields an average wait of $(1/12 \times 1 + 11/12 \times 11)$, or 10 1/6 hours.

The derivation isn't as important as the bottom line: any variation in the buses' arrival times results in a longer waiting time for you even if the average number of buses per day remains the same. The buses aren't abusing you; longer waits are simply what happens with any variation in the bus arrival times.

Longer waiting times occur generally in situations ranging from supermarket lines to doctors' offices. Even though the average number of customers or patients arriving per hour and the average time spent on each should not result in bottlenecks, the expectation that bottlenecks will not occur usually hinges on the evenly spaced arrival of people and the same amount of time being spent on each. When this isn't the case, variation is the reason for your delay, not some cosmic malevolence.

Of course, we should not always count long waiting times as something negative; if we are waiting for obnoxious dinner guests to arrive, for example, long waiting times could be deemed positive. We do seem to have a genius for focusing on the negative. Doing so probably confers survival value, and this may be the true basis for Murphy's Law and the usually misplaced albeit common feeling of being put-upon.

PSYCHOLOGY, PERSPECTIVES, AND PARANOIA

Psychological fragility also helps explain our tendency to feel more aggrieved than aggrieving. There is a remarkable disparity, for example, between the degree of overt approval others need from us in order to feel liked and respected and the degree of private disapproval we can have toward them and continue to like and respect them. Positive and negative appraisals are not on an equal footing.

People's sensitivities and vulnerabilities make frank disapproval difficult. Gossip allows us to express some disapproval, but once again there is a gulf between the spite we usually attribute to others who gossip about us and the evenhandedness we feel we demonstrate when we gossip about others. Confidentially expressing a modicum of ridicule or disparagement toward others is perfectly consistent with our affection and respect for them. Unfortunately, few can tolerate others' (non-family members') ridicule or disparagement and still believe in their affection or respect.

Many political issues arise from that put-upon feeling and its close relative, the feeling of specialness. An interesting exercise is to note which members of Congress are most likely to be described in press reports as being the deciding vote when a bill passes with only a one-vote margin. Mathematically, of course, each member of the majority can be said to have cast the pivotal vote. Psychologically, however, the surprise elicited by a congressman's vote (say, in going against his or her party) would seem to be one factor in having been depicted as the deciding vote. The extent to which the congressman wavers in public would seem to be another, as would the recency of the congressman's decision.

The nice thing about such a close vote from the viewpoint of local media is that dozens of editors and assorted pundits can claim that were it not for their congressman the bill would not have passed. If the congressman hadn't been buttonholed by the president, hadn't had an epiphany, or hadn't received an infusion of "soft" money, he or she might have voted the other way. The one-vote margin does what (local) media usually must do for itself: make social issues personal ones.

More generally, psychology supplies a number of girders that help bridge the gap between statistics and stories. Murphy's Law is just one of a number of psychological roadblocks that often obscure our perception of chance phenomena and therefore, eventually, our view of ourselves. A person who is terrified by every new disease and unimpressed by the hazards of smoking or drunk driving is different (not fundamentally different, but not trivially so either) from someone with a more accurate appreciation of the relative risks we all face. A voluminous psychological literature exists on our perception of risk and chance.

Views on many issues are greatly affected by the ease with which one can retrieve various items from memory, as suggested by answers to the following questions: Do more words have "r" as a first letter or as a third letter? What about the letter "k"? Most people judge incorrectly that more words have these letters in the first position than in the third position since, as Amos Tversky and Daniel Kahnemann have theorized in their classic *Judgement Under Uncertainty*, the former words are more readily available to our memories. Words such as *road, radio, Rembrandt, ratatouille,* and *rheumatism* are easier to dredge up than such words as *abroad, daring, Vermeer, vermicelli,* and *thrombosis.*

A number of refinements and corollaries to the so-called availability error have been identified. Prominent among them are the halo effect—the inclination of people to judge a person or thing in terms of one salient characteristic (say, ivy-league credentials) and the anchoring effect—the tendency of people to accept, or at least to not stray far from, the first number or fact presented to them within a given discussion (say, the number of slaves brought to the Western Hemisphere). Each of these tendencies describes a very common conceptual shortcoming. In moving from the raw givenness of statistics to the malleable interpretations of stories, such tendencies play a pivotal, although largely unrecognized, role.

The sum of the behaviors of many people is less well understood. Crowd behavior sometimes slips into mass hysteria, and the catalyst sometimes seems to be a single individual. Even in organizations much tamer than crowds—organizing committees, for example—as Irvin Janis and other researchers have demonstrated, interactions among members frequently tend to engender bias and grave misestimations of probabilities. Since members wish to be valued by the group, they freely express opinions in line with what they perceive to be the group's attitudes and suppress views that run counter to those of the group. A self-reinforcing, prejudicial breeze soon kicks up. Leaders arise who are more extreme than the average member; they typically pick yes-men rather than more independent-minded members and are deferred to by most others—particularly if said leader can influence the latter's careers.

Investigations also suggest that groups with such leaders are more likely to embark on risky undertakings based on meaningless coincidences and do so with more certitude than are solitary individuals, as was the case with the Heaven's Gate suicides, for example. (Such studies remind me of one of the few biblical injunctions quoted by Bertrand Russell: Follow not a multitude to do evil.) Germane and more encouraging are research results indicating that people who take an unpopular stand in public are much less likely afterward to be swayed by conformist statements and actions than people who express similar ideas in private.

How then should people go about revising their estimates of the likelihood of various events? One answer is Bayes' Theorem. Let me illustrate (without equations) by adapting one of those artificial problems that over the decades have vexed thousands of students and charmed seven or eight. A woman witnesses a break-in in a certain town, 85 percent of whose residents we'll call pink (following the comedian George Carlin's suggestion) and the other 15 percent brown. The witness claims that the burglar was brown, and scientific tests suggest that under the conditions obtaining at the time of the break-in she is correct in her color identifications 80 percent of the time. So given the witness's testimony (and assuming that the town is socioeconomically integrated), what is the probability she is correct that the burglar was brown? Many people will say 80 percent, but the answer is only 41 percent.

The following table clarifies the situation. Assume 100 burglaries take place and that the woman, like a character in a television mystery series, manages to witness every one of them. If 15 of 100 burglars are brown, and the witness is correct 80 percent of the time, then she will likely identify 12 (80 percent of 15) as brown and the other 3 as pink. She will also classify 68 of the 85 pink burglars (80 percent of 85) as pink and the other 17 as brown. So of the 29 burglars she would identify as brown, in fact only 12 are brown. Thus the conditional probability that the burglar is brown given the witness's testimony that the burglar is brown is by Bayes' Theorem 12/29, or only *41 percent!*

The burglar really is	brown /	pink	
The witness says brown	12	17	29
The witness says pink	3	68	71
	15	85	

Bayes' Theorem also helps us understand our natural but unwarranted inclination to vastly overestimate the probability of rare events. Despite press reports, for example, the rape of a child by a parent is relatively uncommon; for the purposes of illustration only, assume that the true incidence is 2 per 1,000 children. If just 1 percent of those not raped by a parent were to misremember or misreport that they were raped, and if 50 percent of those actually raped by a parent were to misremember or misreport that they were not, most of us might believe this would underestimate the real incidence of parental rape, or at least would not significantly overestimate it. We would be wrong. To make the arithmetic easier, assume we survey a random sample of 1,000 people to determine how many were raped by their parents. Since the true incidence (we are assuming) is 2 per 1,000, and since 50 percent of those raped will misremember or report that they were not, the survey is likely to turn up 1 true positive report. Yet of the 998 adults who were not raped, 1 percent—that is, approximately 10—will misremember or misreport that they were raped. Thus, the survey would find that the incidence rate is 11 (1 true report plus 10 false ones) per 1,000, or 5.5 times the true rate of 2 per 1,000.

Also of note is the fact that vivid, widely publicized stories of such crimes are likely to increase, even if only by 1 percent, the percentage of people reporting them.

•   •   •

A different bit of statistics the misinterpretation of which often leads to faulty appraisals of ourselves and others is the phenomenon of regression to the mean. Regression to the mean is the tendency for an extreme value of a random quantity (the values of which depend on a number of factors and which cluster around an average or mean) to be followed by a value closer to the average or mean. Very intelligent people, for example, can be expected to have intelligent offspring, but their offspring usually will not be quite as intelligent. (Why, since the point would be the same, does substituting *stupid* for *intelligent* in the last sentence seem offensive, a regression to meanness as well as to the mean?)

A comparable tendency toward the average holds for restaurants that seem to serve a delectable meal on the first visit—and thereby earn temporary status as one of my wife's favorite restaurants—only to disappoint on a return visit. In this example a lack of parallelism between the restaurant's apparent deterioration and improvement obscures the phenomenon: If the first meal at a restaurant brings on diarrhea, there isn't generally a return visit that would benefit from regression to the mean.

Whether it involves improvement or deterioration, people frequently attribute the regression to the mean to the actions of an agent, rather than to the behavior of any random quantity dependent on many factors. The sequel to a good CD usually is not as good as the original. The reason may not be the avarice of the music industry in exploiting the CD's popularity, but simply regression to the mean. Likewise, a record-breaking year by an athlete will likely be followed by a less impressive year, probably not because he slacked off but because of the tendency to regress to the mean.

We think more in terms of stories and protagonists' deterioration or improvement than in terms of statistics and regression phenomena, but some would rather attribute almost any phenomenon to an agent (whether a Satan, a Santa, or someone in-between) than to workings of chance. A natural partiality to conspiracy stories and the significance of coincidence is hard-wired into us, waiting only for that certain combination of ignorance, events, and stress to flare into the dread conceptual disease PPP (Probabilistically Provoked Paranoia) that we see so frequently.

Our personal viewpoints and psychological foibles also play a role when we choose one of several alternatives based on how they are presented to us. Even in mathematically equivalent cases, we're swayed by artful phrasing. In one investigation subjects generally choose to receive 300 dollars with probability 20 percent rather than receiving 200 dollars with probability 25 percent. This is reasonable since the average gain in the first case is 60 dollars (20 percent of 300), whereas the average gain in the second is only 50 dollars (25 percent of 200). What isn't so reasonable is that subjects generally made the opposite choice when their options were stated in terms of stages.

The alternative framing posits two stages: With a probability of 75 percent the subject is eliminated at the first stage and receives nothing. If someone reaches the second stage, he or she has the option of receiving 200 dollars for certain or 300 dollars with probability 80 percent. This is equivalent to a choice between 200 dollars with probability 25 percent (since 25 percent is 100 percent minus 75 percent) or 300 dollars with probability 20 percent (since 80 percent of 25 percent is 20 percent). In this case, however, the majority choose the seemingly safer 200 dollars option, influenced apparently by the idea of certainty.

In another study subjects generally choose to receive a certain gain of 800 dollars rather than gamble on winning 1,000 dollars with probability 85 percent and winning nothing with probability 15 percent. This occurs even though the second alternative yields 850 dollars on average. Subjects generally make the opposite choice when similar options are stated in terms of losses. If asked to choose between a certain loss of 800 dollars or an 85-percent chance of losing 1,000 dollars and a 15-percent chance of losing nothing, the majority choose the second alternative even though it entails an average loss of 850 dollars.

Other scenarios, many of them matters of life and death, support the proposition that people are considerably more willing to take risks to avoid losses than they are to achieve gains. Not surprisingly, stories and novels dealing with desperadoes in grim circumstances are much more engaging than accounts of wealthy and contented people striving to improve their lot.

Although we may lack lucidity on some of these probabilistic puzzles, rarely are we short of confidence. If Yeats was right when he wrote, "The best lack all conviction, while the worst are full of passionate intensity," then investigations by Boris Fischhoff and others on overconfidence suggest that most of us aren't very admirable. (Interestingly, some studies have indicated that one of the few categories in which American math students are ranked first in international studies is self-confidence.) We're so often cocksure of our decisions, actions, and beliefs because we fail to look for counterexamples, pay no attention to alternative views and their consequences, distort our memories and the evidence, and are seduced by our own explanatory schemes.

This tendency is apparent in the nasty biographies, press exposés, and embarrassing lawsuits now so common. They should come with a warning label stating that the amount of dirt discovered on the subject in question is seldom the most important indicator of his or her worth. The significance of one auxiliary statistic in particular is unappreciated: the *ratio* of the amount of dirt unearthed to the time and resources spent digging for it (or for something that can pass for dirt). As I write news reports state that 30 million dollars have been spent thus far in an effort to uncover alleged wrongdoing by Bill or Hillary Clinton in the Whitewater affair. I don't think I have a particularly disreputable group of friends and acquaintances, but few could withstand a 30-million-dollar investigation into their private lives. (In a later chapter I will return to the question of the intensity with which we look for something and its effect on what we find.)

The way these "investigations" and stories can bias our view of others and of ourselves is illustrated by a classic psychological experiment by Richard Nisbett, whose subjects were told of two specific firemen, one successful, one not. Half the subjects were told that the successful fireman was a risk taker and that the unsuccessful one was not. The other half were told just the opposite. They then were asked to characterize good firemen in general. After doing so, they were informed that the specific firemen did not exist and that the experimenters had simply invented them. Oddly, the subjects continued to be strongly influenced by whatever explanatory profiles they had concocted for themselves. If they had been told that the risk-taking fireman was successful, they continued to think that prospective firemen should be chosen for their willingness to take risks; if they had been told the opposite, then they continued to think that. If asked to account for the connection between risk taking or its absence and successful firefighting, the members of each group gave a cogent explanation consistent with the story they originally told themselves.

More generally, we tend to rationalize coincidences and stories of all sorts. We try to force them to make sense, and we may even try to rescind the great and pervasive Law of Unintended Consequences (of which Murphy's Law is a special case). Because the stories we believe become, at least metaphorically, a part of us, we are disposed, perhaps out of a sense of self-preservation, to look always for their confirmation, seldom their disconfirmation. (I'm tempted to say that even a cursory glance around us confirms this.)

One very simple notion from statistics, were it used regularly by people, would go a long way toward minimizing the blinding effects of this natural tendency and insuring more critical thinking about confirmation and disconfirmation. The notion is a so-called two-by-two table in which the frequencies of all four possible relations between any two simple phenomena, A and B, are examined: not just A and B, which usually strikes us first, but also A and not B, not A and B, and not A and not B. The idea is so elementary it can be taught to young children and career politicians.

THE BIBLE CODE AND SEX SCANDALS

There is another psychological foible that ought to be mentioned. In trying to make sense of things, people are much more likely to attribute an event to an agent's will than to chance if the event has momentous implications. In one experiment, for example, a group of subjects is told that a man has parked his car on an incline and that subsequently it has rolled down into a fire hydrant. Another group is told that the car has rolled into and injured a pedestrian. The members of the first group generally view the event as an accident; the second group is more likely to hold the driver responsible. Other studies confirm that the more emotionally fraught an event or phenomenon is, the more eagerly people search for a story to make sense of it.

This psychological tendency, together with our natural preference for confirmation over disconfirmation and our fascination with coincidence, help explain the appeal of the Bible codes mentioned in the Introduction. Although not a "straight" account, the following very mild parody attempts to elucidate the interpretation of the probabilities involved in the controversy and indicate what they do or do not mean.

•   •   •

The brouhaha involving the codes to be found in various holy books brings to mind a lesser-known recent discovery that has been suppressed by President Clinton's lawyers. Encoded within the U.S. Constitution is a prophecy of the Lewinsky sex scandal! Placed there presumably by the Founding Fathers, the 10 letters in the words *Bill* and *Monica* appear sequentially at regular intervals within the revered historical document. Remarkably similar to those of some of the biblical codes, the details are instructive: the interval between the successive letters in *Bill/Monica* is 76; that is, at a particular position within the Constitution, there is a *b*, followed after 76 letters by an *i*, followed after another 76 letters by an *l*, and so on until the *a* of *Monica* is reached 76 letters after the *c*. (These are the ELS, equidistant letter sequences, that have attracted so much attention.)

Upon discovery of this seemingly prescient sequence of letters it is only natural to wonder about the probability of its occurrence. If we assume as a first approximation that the letters of the Constitution are randomly distributed, the probability of observing the 10 letters in *Bill/Monica* in any given set of 10 equidistant letter positions within the Constitution is easily computed: simply multiply the probabilities of occurrence of each of the 10 letters in the sequence. (If, for example, in any given position the probability of *b* is .014, that of *i* .065, and that of *l* .011, then the probability that the 4 letters in *Bill* appear in any 4 given positions is .014 × .065 × .011 × .011.) Thus the product of these 10 small numbers—let's call it *P*—is a truly infinitesimal probability.

Because this likelihood is so minuscule, we might think that the occurrence of a *Bill/Monica* sequence at some particular set of positions within the Constitution is an extraordinary event; but we must be careful about our understanding of this extreme improbability. The meaning is this: If we were to choose one text from the collection of all texts with the same number of letters of each kind as the U.S. Constitution, and if we were to designate an ordered list of 10 particular letter positions and check to see whether the letters in *Bill/Monica* were in these designated positions, the probability is *P* that they would be.

Such a procedure does not, however, reflect the way in which the *Bill/Monica* sequence in the Constitution was discovered. In our probability calculation we assumed that the letter sequence and positions were specified *beforehand* and the text selected and observed *afterward*. In the actual discovery of the Constitution code the observation came first—that is, the *Bill/Monica* sequence of letters was found in the document by, we can imagine, a computer-savvy scholar in some think tank on the Potomac. Once the sequence was found the question of the likelihood of its occurrence became moot.

Equally salient is that the *Bill/Monica* ELS need not occur in some particular place in the Constitution. We are not especially concerned that the sequence begin at, say, the 14,968th letter; rather we look for this pattern beginning *anywhere* in the Constitution—that is, we look at all the many different letter positions in which the 76-letter equidistant pattern can begin (assume there are *X* such letter positions within the Constitution) to see if we can find at least one instance of it. The probability of observing the *Bill/Monica* pattern for this procedure is considerably larger, roughly equal to $P \times X$.

Now suppose we do not search merely for an interval of 76 between the letters in *Bill* and *Paula*, but rather search for the pattern at all possible intervals between, say, 1 and 1,000 and beginning anywhere in the Constitution. With this procedure the numbers change again. The probability that we observe the pattern *Bill/Paula* is approximately equal to $P \times X \times 1,000$, and this number is not so amazingly small.

We can again increase the probability of finding such a sequence by further expanding the number of ways in which it might occur. We might allow backward searches, or look along diagonal lines in the text, or (as is the case with the Bible codes) permit distinct ELSs for *Bill* and *Monica* to be nearby but separated in the text, or search for alternative names for the president or his paramour(s), or loosen the constraints in indefinitely many other ways.

If our search for these sequences is not conducted openly, and if cases in which nothing appropriate is found are discarded (e.g., nearby ELSs for *zucchini* and *squash*), and if we go public only with the interesting sequences we do find and compute probabilities in a simplistic way, then it is clear that these sequences do not mean what they may seem to mean on the surface. Performing a procedure one way and computing a probability associated with a different procedure is, to put it mildly, just not kosher.

Almost all of the many biblical codes—whether from Jewish, Christian, Islamic, or modern sources, whether utilizing Cabala or Monica—have defects vaguely similar to those of the Constitutional codes.* The statistical pa-

---

*As far as I know, the *Bill/Monica* equidistant letter sequence does not appear in the U.S. Constitution, although researchers Brendan McKay and David Thomas have located surprising sequences virtually everywhere they have searched: in Genesis (which contains more than

per mentioned in the Introduction may also illustrate a different, more subtle defect having to do with unintentional biases in the choice of sought-after sequences, vaguely defined procedures, the variety and contingencies of ancient Hebrew spelling and references, or Ramsey's Theorem (to be discussed in a later chapter)—a deep mathematical result about the inevitability of order in any sufficiently long sequence of symbols. This in fact was the reason for the paper's publication, not a belief in coded prophecies. Common sense underscores the inanity of basing any political, spiritual, or sexual judgments on these contextless numerological oddities.

•   •   •

The Bible codes are just one of the latest manifestations of our natural tendency to read significance into coincidence. I've written in *Innumeracy* and *A Mathematician Reads the Newspaper* of the stunning insignificance of the vast majority of coincidences. That 2 people in a group

---

two dozen ELSs for Hitler and Stalin), Shakespeare, Tolstoy, Melville's *Moby-Dick* (which contains an ELS for "Oceans hold joy"), a U.S. Supreme Court ruling on creationism, and *Chicago Tribune* editorials. Thomas has even found that the following verse from the King James Version of Genesis contains a short, topical ELS for *Roswell*, which is capitalized, and, beginning with the underscored *u* of *thou*, one for *UFO* as well:

And hast not suffered me to kiss my sons and my
daughteRs? ThO<u>u</u> haSt noW donE fooLishLy in s<u>o</u> doing.

The likelihood of each of these specific ELSs is minuscule, from which nothing can be inferred except that it is easy to misconstrue tiny probabilities.

share a birthday, for example, is quite likely if the group numbers 30 or more. If there are 23 people in the group the probability is 1/2 that at least 2 share a natal day. And if 2 people, A and B, are chosen at random from national phone books, the probability is overwhelming that they will be linked by 2 intermediates: that is, person A will know someone who knows someone who knows person B (although A and B will not likely know of the intermediate links). Seemingly uncanny lists of similarities exist between Presidents Lincoln and Kennedy. I say "seemingly" because comparably long lists exist for Presidents McKinley and Garfield as well. The first letters of the planets in order of their distance from the sun are MVEMJ*SUN*P. The first letters of the months are JFMAMJ*JASON*D. That some predictions by psychics will come true is to be expected on grounds of probability alone. Et cetera, et cetera, et cetera.

World-class coincidences are sometimes termed miracles—with the proviso that the result be positive. (It's generally not termed a miracle when a rare earthquake levels a building on the only day of the year it is full of schoolchildren.) As with more mundane versions, the vast majority of "miracles" mean nothing; some point to valuable yet overlooked connections; and a few suggest a void in our understanding. David Hume's insight about this latter variety of miracle is too little appreciated. Hume observed that every piece of evidence for a miraculous coincidence—that is, for a violation of natural law—is also evidence for the proposition that the regularities the alleged miracle violated are not laws of nature after all.

The most amazing coincidence of all would be the complete absence of all coincidences. This statement is a rough rephrasing of the gist of the aforementioned theorem by the English mathematician Frank Ramsey and his intellectual heirs (I reiterate most of these points despite the fact that repetition of nonsense is much better tolerated than repetition of its debunking, which generally comes across as scolding and earnest).

One consequence of the mistaken belief that coincidences are special and almost always significant is their rarity in most modern fiction, where introducing one is considered a form of cheating. We've moved *too* far away from Victorian novelists, who regularly inserted outlandish coincidences into their works. If Charlotte Brontë stretched the long arm of coincidence to the breaking point, as was once remarked, most modern writers have reduced it to an unnatural stub incapable of reaching out to a larger world. Coincidences are the ubiquitous stuff of life and leaving them out of a novel or movie makes plot and character development necessarily more deterministic and less lifelike.

Some Modernist forms of literature make a conscious attempt to reflect the aleatoric nature of life, and these stream-of-consciousness, fragmented, collage-like works do, like newspapers, contain many coincidences. In comic novels too a greater tolerance for coincidence exists, in part for the comedy they generate. This greater tolerance also may be a reflection of the cooler gaze of the comic author who is more willing to rip events out of meaningful contexts within characters' lives and point out their coincidental connection. (The allowability of such ripping out of events and entities and substituting them into other contexts differs considerably according to discipline, as we'll see in the next chapter. In probability and in mathematics generally, substitution of equals for equals is allowed; in storytelling, especially in the first person, substitution is usually improper.)

Whether we are comfortable with the insignificance of most coincidences or insist on always finding a Meaning behind them is, in the end, a critical and revealing aspect of our personalities and world outlooks.

A TRICKY TALE AND PROBABILISTIC COUPLING

Some coincidences are significant, but not for ostensible reasons. A magic trick of recent vintage nicely illustrates a way in which two people can "cognitively couple" and generate an otherwise hard to explain coincidental merging. This example has relevance as well to biblical codes.

The card trick was invented about fifteen years ago by the mathematician Martin Kruskal and can be explained most easily in terms of a deck of cards with all of the face cards removed. Imagine two players, Tricked and Trickster. Trickster asks Tricked to pick a secret number—say X—between 1 and 10, and goes on to instruct Tricked to watch for the Xth card as Trickster slowly and one by one turns over the cards in a well-shuffled deck. When the Xth card is reached—say it's a Y—it becomes Tricked's new secret number and he is asked to watch for the Yth succeeding card after it as Trickster continues to slowly turn over the cards one by one. When the Yth succeeding card turns up, its value—say Z—becomes Tricked's new secret number, and again he is asked to watch for the Zth card succeeding after it for his new secret number, and so on.

Thus if Tricked first picks 7 as his secret number, he would watch for the 7th card as Trickster slowly turns over the cards. If the 7th card is a 5, his new secret number would become a 5, and he would watch for the 5th card after it. If the 5th card after this is a 10, 10 would become his new secret number and he would watch for the 10th card after it to determine his new secret number. As they near the end of the deck, Trickster turns over a card and announces, "This is your present secret number," and he is almost always correct. The deck is not marked or ordered, there are no confederates, there is no sleight of hand, and there is no careful observation of Tricked's reactions as he watches the cards being turned over. How does Trickster accomplish this feat?

The answer is cute. At the beginning of the trick Trickster picks his own secret number. He then follows the same instructions he has given to Tricked. If he picked 3 as his secret number, he watches for the 3rd card and notes its value—say 9—which becomes his new secret number. He then looks for the 9th card after it—say a 4—and that becomes his new secret number.

Even though there is only 1 chance in 10 that Trickster's original secret number is the same as Tricked's original secret number, it is reasonable to assume and can be proved that sooner or later their secret numbers will coincide—that is, if two more or less random sequences of secret numbers between 1 and 10 are selected, sooner or later they will, simply by chance, lead to the same card. Furthermore, from that point on the secret numbers will be identical, since both Tricked and Trickster are using the same rule to generate new secret numbers from old. Thus all Trickster does is wait until he nears the end of the deck and then turn over the card corresponding to his last secret number, confident that by that point it probably will be Tricked's secret number as well.

Aside from the pleasures of understanding it, does this trick have any real-world analogues? Note that the trick works just as well with more than one Tricked person or even with no Trickster at all (as long as the cards are turned over one by one by someone). With a large number of people, each picks his or her own initial secret number and generates a new one from the old one in accordance with the abovementioned procedure; all eventually will have the same secret number and thereafter will move in lockstep.

If we allow people's new secret number to be determined in a more complicated way from several of its predecessor secret numbers instead of just from its immediate predecessor, and if we change the scenario from turning cards over one by one to some other sequential and numerical activity such as the lottery or the stock market, we see the potential for lockstep behavior on a large scale to develop naturally. If many investors, for example, use the same computer software (i.e., with the same rules for determining when to buy or sell) it is conceivable that some attenuated variant of large-scale lockstep behavior might result whatever the investors' initial positions.

I propose the following religious hoax. Consider a holy book with the compelling property that no matter what word from the early part of the book is chosen the following procedure always leads to the same climactic and especially sacred word: Begin with whatever word you like; count the letters in it; say this number is X; proceed forward X words to another word; count the letters in it; say this number is Y; proceed forward Y words to another word; count the letters in it; say this number is Z; iterate this procedure until the climactic and especially sacred word is reached. It's not too hard to imagine frenzied checking of this procedure using word after word from the early part of the holy book and the increasing certainty that divine inspiration is the only explanation for such a phenomenon. If the generating rule were more complicated than the simple one used in this example, the effect would be even more mysterious.

BAYES' THEOREM AND
REVISING OUR STORIES

All of us informally estimate probabilities every day, and most of us, whether as individuals or in collaboration with others, frequently revise these estimates. Before getting to its revision, however, something should be said about the theoretical definition of probability.

Unfortunately, there remains considerable controversy over the meaning of probability. Some have conceived of it as a logical relation, as if one could just glance at a die, note its symmetry, and decide by logic alone that the probability of a 3 turning up must be 1/6. According to others, relative frequency is the key to the analysis, the probability of an event being a shorthand way of indicating the long-run percentage of the time it occurs, *long run* usually remaining unexplained. Still others suggest that probability is a matter of subjective belief and is nothing more than an expression of personal opinion growing out of plausible stories and daily experiences.

Although the debate still simmers, mathematicians—like losing generals everywhere—have simultaneously retreated and declared victory. They have observed that since by all reasonable definitions probability ends up possessing certain formal properties, probability simply should be defined as whatever satisfies these formal properties. Such a definition may not be philosophically gratifying, but it is mathematically liberating and serves to bring into some minimal accord the disparate accounts of the notion.

The properties that all accounts of probability seem to possess were listed by the Russian mathematician A. N. Kolmogorov and capture such elementary understandings as the following: The probability of an event occurring, measured by some number between 0 and 1 (or, equivalently, by some percentage between 0 percent and 100 percent), should be equal to 1 (or 100 percent) minus the probability of the event not occurring. The probability of one of several mutually exclusive events occurring (say a 1, 3, or 5 turning up on a die) is the sum of the probabilities of each of the events (1/6 + 1/6 + 1/6). The probability of several independent events all occurring is the product of their respective probabilities. These and other properties or axioms are not the topic of this book, but I would like to examine an important consequence of them.

Conditional probabilities are probabilities in the light of, or given, certain evidence. The probability of a randomly chosen adult weighing less than 130 pounds is, let's assume, 25 percent. The conditional probability that someone weighs less than 130 pounds given that he or she is over 6 feet 4 inches tall is, I would estimate, much smaller than 5 percent. Note also that the conditional probability that one can speak Spanish given that one is a citizen of Spain is, let us say, approximately 95 percent, whereas the conditional probability that one is a Spanish citizen given that one can speak Spanish may be less than 10 percent.

Bayes' Theorem is a formula that tells us how we should modify our sometimes subjective conditional probabilities, and hence, indirectly, how we should alter the stories that give them context and meaning. The network of conditional probabilities each of us assigns in idiosyncratic ways to myriad events and hypotheses is quite intricate. We differ in the way we attach rough probabilities to happenings, and differ even more in the probabilities we assign to the associations between happenings.

This tangled network of probability estimates, propensities, and beliefs is, in a sense, a map of our minds and dynamically interacts with new experiences and old stories that we constantly edit. This network also depends, as we'll see in the next chapter, on other people's probability estimates, propensities, and beliefs. Revision of our subjective conditional probabilities generally brings our personal viewpoint—no matter how unreasonably idiosyncratic—into better agreement with new, more objective evidence. Thus, despite our immunity to facts on occasion, conditional probabilities are a crucial link between the impersonal world and the stories we tell about ourselves and others.

What exactly is Bayes' Theorem? Although we each use it implicitly in revising our probability estimates, only statisticians regularly use the formal version. For the record and without formal notation, Bayes' Theorem states that the conditional probability of a hypothesis given some new piece of evidence is equal to the product of (a) the initial probability of the hypothesis before the evidence and (b) the conditional probability of the evidence given the hypothesis, divided by (c) the probability of the new evidence. The formula and its derivation are not important here, since often it is easier to construct tables (as in the preceding burglar example) than to derive or use the formula. What is important is that Bayes' Theorem provides us with a way to incorporate new, objective information into our personal, subjective outlooks. Unfortunately, it can sometimes lead to correct but counterintuitive results, especially in situations involving individuals or rare events.

Another problem with Bayes' Theorem is that real-life probabilities may be imbedded in many different ways in countless stories and arguments, and new evidence therefore may be filtered and factored into the Bayes probability revision machine in many (sometimes incommensurable) ways as well. Everyday stories are always nebulous and multifaceted. Some people, for example, who learn that a lawyer's last 5 clients were convicted might revise downward their estimate of their chances for acquittal were they to select him as their attorney. Others, giving greater weight to the fact that all the clients were wealthy and came from different parts of the country, might see the lawyer as an eminent practitioner who takes on only the most difficult cases, and therefore revise their estimate upward.

More nuanced legal cases provide better illustration.

COMPLEX LEGAL STORIES AND INFERENCE NETWORKS

Probability and law have an uneasy relationship. Quantifiable probability and qualitative plausibility, despite Bayes' Theorem, cannot always be reconciled; attaching a numerical value to a plausible argument often is simply wrong-headed. Probabilists are sometimes reductionist and impatient with the subtleties and protections of law, while lawyers are sometimes innumerate or dismissive of the insights afforded by probability. Even so, each field contains very good examples of notions from the other. A proposition that is rejected only if found to be quite improbable, the null hypothesis of statistical practice is exemplified nicely in the legal presumption of innocence.

Some recent celebrated cases in which transcripts run thousands of pages might be clarified (for interested onlookers, if not always for jurors) if the argument was arrayed in a chart. In such a chart various bits of evidence and inferential steps leading to the conclusion of guilt

would be laid out, organized, and logically chained together. Each intermediate step in the argument also
would have its own subsidiary chain of evidence (brought
out in cross-examination and redirect examination) supporting or undermining it. Some of the steps in these
subsidiary chains would have subsidiary chains of their
own confirming or disconfirming them. (The tree diagrams mentioned in the last chapter are relevant here.)
Now *if* the jurors could assign elementary probabilities to
these bits of physical evidence and testimony, then repeated application of the laws of probability, in particular
of Bayes' Theorem, would yield a likelihood for an overall conclusion of guilt. If this overall probability were not
sufficiently high, the defendant would be declared not
guilty.

Such a procedure, here somewhat simplistically described, was put forward by John H. Wigmore, a former
dean of Northwestern University's Law School, in his
1937 book *Judicial Proof: As Given by Logic, Psychology, and
General Experience and Illustrated in Judicial Trials.* Joseph B.
Kadane and David A. Schum apply the method at great
length in their book, *A Probabilistic Analysis of the Sacco and
Vanzetti Evidence.*

The probabilities different jurors or onlookers assign to
the veracity, objectivity, and observational sensitivity of witnesses will vary depending on their views and life experiences. Some are truthful but obtuse. Some are biased but
acute. The probabilities jurors assign to the credibility, relevance, and cogency of pieces of evidence will vary for the
same reasons. One piece might be indubitable, but irrelevant. Another might be of compelling relevance, but ques-

tionable. Distinguishing these characteristics is therefore essential when determining probabilities. Some probabilities will be more or less constant for every observer; the probability that the murderer was at the scene of the shooting, for example, will be adjudged to be 1 by everyone. Different stories will establish different linkages among the witnesses' testimony and tangible evidence.

In addition to personal variations in probability assessment are psychological illusions similar to those above-mentioned, to which we are all subject. Studies have shown, for example, that an action by a defendant that raised the likelihood of an accident from 0 in 1,000 to 1 in 1,000 generally will be viewed as much more reckless than an action by a different defendant that raised the likelihood of an accident from 5 in 1,000 to 6 in 1,000, or even one that raised the likelihood from 5 in 1,000 to 10 in 1,000.

The evidential, probabilistic, and testimonial elements of complicated cases are so overwhelmingly complex that artificial means of organizing the information and drawing coherent inferences from it can be extremely useful. Such means can take the form of hand-drawn Wigmore charts (formally described in Wigmore's book) or of software packages (such as ERGO) designed specifically for navigating through such inference networks. I reiterate that jurors and others can assign whatever probabilities they like to bits of evidence and testimony and to links in various contending stories of the crime. The charts or software merely insure that these possibly absurd assignments and verdicts derived from them will be internally consistent—they do not guarantee more than that.

O. J. Simpson and
the Crime of Statisticide

Masochistic readers may try to envision the massive Wigmore charts needed to map the O. J. Simpson case, which is more than tangentially germane to Bayes' Theorem, coincidences, and the relations between individual viewpoints and societal norms. The following discussion is adapted from an Op-Ed I wrote for the *Philadelphia Inquirer* after the first verdict:

Adding to the dissatisfaction with the Simpson saga were numerous instances of what might be termed *statisticide.* Let me begin with a refrain constantly repeated by attorney Alan Dershowitz during the trial. He declared that since fewer than 1 in 1,000 women who are abused by their mates go on to be killed by them, the spousal abuse in the Simpsons' marriage was irrelevant to the case. Although the figures are correct, Mr. Dershowitz's claim is a stunning non sequitur; it ignores an obvious fact: Nicole Simpson was killed. Given certain reasonable factual assumptions on murder and spousal abuse, it can be easily shown using the Bayes' Theorem that if a man abuses his wife or girlfriend and she is later murdered, the batterer is the murderer more than 80 percent of the time (a nice demonstration of this by Jon Merz and Jonathan Caulkins appeared in a recent issue of *Chance* magazine). Thus, *without any further evidence,* there was mathematical warrant for immediate police suspicion of Mr. Simpson. I'm certainly not advocating the abrogation of our Fourth Amendment rights; I am merely pointing out that looking to Mr. Simpson was not, on the face of it, unreasonable, nor was it an instance of racism.

Another mathematical notion that could have been used more trenchantly during the trial is that of statistical independence. As explained earlier, two events are statistically independent if one's occurring does not affect the likelihood of the other's occurring. Furthermore, when two happenings are independent (e.g., multiple coin flips), the probability of their both occurring is simply the product of their respective probabilities.

If the various bits of incriminating evidence were independent manifestations (and many were), then their respective probabilities should be multiplied to obtain the probability of all of them turning up. Forget DNA for the moment and consider only the probabilities of two simple physical findings. What is the likelihood that the perpetrator's footprints leading from the crime scene would be size 12, Mr. Simpson's size? And what is the likelihood that Mr. Simpson would sustain a cut on the left side of his body the very night the murderer did (judging from the blood spots to the left of the footprints)? Estimates for these probabilities may vary, but let's be generous and say that they are as high as 1 in 15 and 1 in 1,000, respectively. The probability of both independent pieces of evidence turning up is the product of their probabilities: one in 15,000, a very strong indicator of guilt *completely separate* from the overwhelming DNA evidence. Bringing in the many other bits of evidence further reduces this tiny probability.

Independence played a role in the DNA testimony as well, where citations of probabilities smaller than 1 in 5.7 billion, the earth's population, were ipso facto viewed by many as instances of prosecutorial exaggeration. But the earth's population has nothing to do with the matter. Since there are incomparably more potential DNA patterns (just as there are incomparably more bridge hands) than there are people on Earth, it makes perfectly good sense to assert that the probability that someone has a particular fragment of DNA (or bridge hand) is 1 in 75 billion (or 1 in 600 billion for bridge). Such minuscule probabilities are the result of multiplying many small probabilities together.

Of course, exculpatory probabilistic arguments also can be made. It can be contended, for example, that the crucial question in the trial was not the probability of an innocent person's having all this evidence arrayed against him, but the probability of a person with all this evidence arrayed against him being innocent, which is quite a different thing.* In the Simpson case, however, this is not a very promising tack, and so the defense was left with their theory of conspiracy and coverup. Attaching a precise probability to such a scenario is not possible, but accepting it entails believing that the same bungling police department that cavalierly ignored all of Nicole Simpson's previous calls for help could and would, upon discover-

---

*An example helps. Imagine a city of approximately one million people, a heinous murder has been committed, and the only evidence available indicates that the murderer has a very rare sort of mustache. Assume further that only two residents of the city have such mustaches. One of these people is innocent, the other guilty. Then the probability that an innocent person has this rare form of mustache is one in a million; one out of the one million innocent people has such a mustache. By contrast, the probability that a person having such a mustache is innocent is *one in two!* Circumstantial evidence, motive, and further physical evidence therefore always should be sought to bolster any single piece of forensic evidence.

ing her death, instantly and without direction devise an elaborate frame-up of Mr. Simpson. Police officers, lab technicians, and criminalists (about whom, with one exception, no evidence of moral depravity was presented) would have to have been implicated in a complicated network of villainy.

Would a Wigmore chart have been any use in the trial, especially given the diagram's necessary complexity? Jurors apparently gave inordinate weight to certain pieces of evidence (the glove not easily fitting) and virtually ignored other pieces (the blood evidence). Nevertheless, by forcing internal consistency such a chart might have encouraged more systematic deliberation. More can be said, but many feel statistics are boring and that we should therefore go easy on the crime of statisticide. Perhaps we should, but not when doing so leads to the pardoning of homicide. There must be some accommodation between seductive narratives and bloody probabilities.

*   *   *

Subjective viewpoints and objective probabilities have at times a troubled relationship that resembles that between informal discourse and formal logic. Applications of probability and statistics require a story, a context, or an argument in order to make sense. As we'll see in the next chapter, however, the logic of stories, everyday conversations, and informal arguments is not always compatible with the formal logic of science, mathematics, and statistics.

3

βetweeɲ ɪnforɱal ðiscourse and ʟogic

Why is it no one ever sent me yet
One perfect limousine, do you suppose?
Ah no, it's always just my luck to get
One perfect rose.

—Dorothy Parker

WHILE AN UNDERGRADUATE at the University of Wisconsin I took a graduate course in mathematical analysis based on an elegant book by Professor Walter Rudin. Although theoretically self-contained, the book was quite abstract.* Had I not developed certain common intuitions and not taken an earlier course in advanced calculus based on another beautiful but more concrete book by Rudin, I wouldn't have had much feel for what was going on in the course. I suspect, however, that I would have done about as well with the technical exercises and formal proofs.

---

*Aside to insiders: Rudin's *Real and Complex Analysis* approached the subjects of integration and measure theory, for example, via the theory of linear functionals.

My point is that formal facility is not the same thing as intuitive understanding. Being able to manipulate symbols and objects like a sidewalk card shark does not necessarily imply any understanding of underlying mathematical principles. Nor does such understanding imply manipulative ability. The fastest solvers of the Rubik cube are usually innocent of the algebraic group theory underlying its solution, while group theorists attempting to solve the cube usually succeed only in giving the illusion of being severely arthritic. One can sometimes get by technically (as card sharks and Rubik cube solvers do) without a conceptual grasp of the relevant notions. More commonly, each facility helps extend the other, with intuition generally coming first and grounding the more technical skills.

This relation between intuitive understanding and formal mathematics holds more generally. Just as statistical notions developed in response to everyday situations, so too have logical arguments and techniques grown out of informal discourse. Inevitable disagreements and natural desires to further their views would favor those individuals who acquired some rudimentary sense of logic and mathematics. Whatever the evolutionary and cultural details, we learned over time to produce not only observations and conversations but also theorems and corollaries. Logical, statistical, and mathematical ideas ripen and eventually achieve a life of their own independent of us, yet these humble foundations and intuitions continue to ground our understanding of them. (This is certainly not to say, however, that our logical, statistical, and mathematical intuitions are always correct.)

We may even regard the history of mathematics as a series of focused conversations about sometimes rarefied ideas. All of the major strides in mathematics, as well as most of the baby steps, are imbedded in historical stories that provide technical advances with an intuitive framework, motivation, extra-mathematical meaning, and significance.

The history of mathematics is, in this sense, not much different from the history of other fields. It is, of course, a history of theorems, but it is also a history of their encompassing narratives and folklore: the Pythagorean legends, the development of our number system, the Arab progress in algebra, the evolution of calculus from Isaac Newton to Leonhard Euler, non-Euclidean geometry, Galois theory and the abstract turn in algebra, Cauchy's insights in analysis, Cantor's set theory, Karl Pearson's statistical tests, Gödel's incompleteness results, A. N. Kolmogorov's axiomatization of probability, the saga of Fermat's last theorem culminating in Andrew Wiles's recent proof of it, and many, many others.

Formal proof and careful computation are usually taken as definitive of mathematics, but as necessary as they are, most students of mathematics (including professional mathematicians) most of the time do not want either. They want what people in other fields want: informal discourse, heuristic explanations, the historical background of the "conversation," connections, intuitions, applications, and answers to the questions, What's the logic here? What's the story? Too often the unsatisfying answers provided them are an unintegrated collection of techniques, algorithms, and tricks.

INFORMAL LOGIC AND US

This book is not concerned with the history of great theorems, but with bridging, or at least clarifying, some of the gaps between formal mathematics and its applications. Here too the encompassing stories and implicit understandings are important. Indeed, with applications such informal discourse is even more critical. For example, applying probability and statistics is much more a matter of grasping the situation, constructing informal arguments, and building comprehensive narratives than of substituting numbers into formulas. Moreover, a broader understanding of mathematical (especially statistical) applications helps narrow the putative gap between stories and mathematics (especially statistics).

No matter how accurate the relevant statistics on affirmative action, for example, they must be imbedded in a story and help advance an argument to be truly meaningful. Moreover, the story and argument are more likely to be flawed and subject to criticism than are the raw numbers. Thus, as Arnold Barnett has observed, it would be a blunder to confuse someone's claim that "I would have been hired if I were a member of that minority group" with the much less credible "I wasn't hired because I wasn't a member of that group." And no matter how high an HMO's subscriber-satisfaction statistics are, the relevant and revealing sample is not a random sample of all subscribers but only of those who are very sick. (Likewise, freedom of expression should not be measured by how likely the average person is to be silenced, but by how likely someone with something to say is.)

In such applications of statistical ideas the story and its everyday logic are primary, the formal statistics secondary. But here (a multimedia trumpet sounds and reverberates; check this book's binding if your copy is silent) is the rub: the everyday logic of stories isn't quite the same as the standard logic of mathematical proof and scientific demonstration. In sharing intuitions, telling stories, and having conversations people often select and interpret what they see in disparate and idiosyncratic ways. People are far more likely to imagine various more or less plausible scenarios, use metaphor and analogy, imbed matters in a specific context, and adopt a particular point of view than they are to look for proofs, studies, and calculations. As a result, the informal logic used in storytelling, conversation, and daily argumentation is more nebulous, inclusive, and people-centered than the standard sort used in scientific deductions and constructing mathematical proofs.

In everyday "story logic" how *we*, the storytellers, characterize an event or person is often critical; our perspective and our ability to "see as" are essential. If a man touches his hand to his eyebrow, for example, we may see this as an indication he has a headache. We may also see the gesture as a signal from a baseball coach to the batter. Then again we may infer that the man is trying to hide his anxiety by appearing nonchalant; that it is simply a habit of his; that he is worried about getting dust in his eye; or indefinitely many other things depending on indefinitely many perspectives we may have and on the indefinitely many human contexts in which we find ourselves. A similar open-endedness characterizes the use of probability and statistics in surveys and studies.

This dependency on personal perspective (even when it is objectively laughable) and particular contexts has led to a great deal of fruitless controversy over the "social construction of reality," most of which is moot when the reality being socially constructed is social rather than physical. It seems incontrovertible that the reality of sporting contests, stock markets, fashions and fads, elections, laws, economic understandings, traffic regulations, and the IRS is to an overwhelming extent socially constructed, and, of course, that the reality of plants, planets, and Planck's constant is not. Mathematics, as may be inferred, is a special case: numbers and theorems exist independently of us, not so their primitive origins, applications, and interpretations. Our relationship with man-made rules, as opposed to scientific laws and mathematical theorems, is somewhat similar to that of people addicted to placebos—we want, believe, and require them to work, and so they do.

Rather than continue on in this woolly postmodernist vein (although "woolliness" is sometimes undervalued, especially by mathematicians), let me describe a curiosity that hints at what can happen at the indistinct border between rigid mathematical structures and those allowing even a small bit of human choice.

WILD CARDS IN POKER AND LIFE

Puzzles involving playing cards can be tiresome (many silently dread the threat "let me just show you a few more"), but Steve Gadbois, John Emert, and Dale Umbach recently rediscovered a fact about the game of poker that should be of interest to people who never have played cards in their lives. Professional gambler and author John Scarne first mentioned it in his book *Scarne on Cards.*

Imagine that you are playing a game of Five-card Draw with two wild cards; this is standard poker in which two cards are designated as taking on any value you decide. Due to these wild cards you and your opponents have some freedom in choosing what hands to pursue and how to declare them.

You still choose the highest possible hand in the usual ordering of hands, the order being determined by the probabilities of getting these hands. The less probable the hand, the higher its rank. Three of a kind is less probable (and hence of a higher rank) than two pair, which in turn is less probable than one pair. These authors observed, however, that the introduction of wild cards and the discretion that they allow players can jumble the order of the various possible hands.

With two wild cards it becomes *more* likely that you will be dealt three of a kind than two pair. (Any pair combined with a wild card is three of a kind.) Since in this situation you are *less* likely to obtain two pair, such a hand should beat three of a kind. Suppose you change the rules and declare this by fiat, so that players choosing between two pair and three of a kind will now choose two pair. Under these altered rules it again becomes *more likely* that they will be dealt two pair rather than three of a kind.

To reiterate: three of a kind is less probable than two pair. The introduction of wild cards makes it more proba-

ble. If two pair therefore is declared a higher hand to reflect its lower probability, it again becomes more probable. The order of other hands as well gets scrambled in
this same irreparable way by introducing wild cards. In
fact, with two wild cards it becomes more likely that you
will obtain one pair than that you will get a bust hand,
and thus a bust hand should beat a pair!

When you have such wild cards there is no way to
achieve what is called a linear order. You cannot rank the
hands according to their probabilities of occurrence as
you can when no wild cards are used—a fairly essential
part of the game to have to renounce.

What might such an esoteric factoid mean for the rest
of us? If the rigidly ordered realm of poker gives rise to
nonorderable outcomes by merely introducing a couple
of wild cards, it is hard to resist the implication that other,
less well-defined domains of endeavor are even more subject to indeterminacy and personal choice. Many of our
life's efforts (marrying, raising kids, conversing, making a
living, learning, buying, investing) are governed by rules
that are gamelike but also at times quite nebulous. In addition to rules, laws, customs, and understandings, there
are exceptions, bluffs, and in effect, wild cards.

Some positive outcome is deemed relatively improbable and hence valuable, and this brings about a greater
striving for its attainment, whether by hook, crook, or
wild card. Greater striving increases the probability of the
outcome's occurrence and hence makes it less valuable.
There is no way to definitively order the possible outcomes; each tentative ordering in effect jumbles the
ranks and brings about a reversal of value for some of the

outcomes. Clearly the element of faddishness, personal choice, and social definition in everyday life (not to mention the stock market) is incomparably larger than in poker, and one can avoid it only by eschewing wild cards or their functional equivalents. This is not easy and probably is not even possible. Life, as the pompous mathematician says, is full of wild cards whose values we determine.

Rules, Substitution, and Probability

How we see and characterize events and people often is an open-ended choice and helps to shape further choices we might make. Someone scoring in the 34th percentile of those taking an exam may blithely choose to describe his performance as putting him in the middle third of those taking it, or he may say that it clearly indicates he has no talent for the subject. Whatever choice he makes closes some doors and opens others. Numbers and facts filtered through our experiences and beliefs become somewhat plastic, and the notions of extensionality and intensionality are useful in clarifying this plasticity.

Standard scientific and mathematical logic is termed *extensional* since objects and sets are determined by their extensions (i.e., by their members). That is, entities are the same if they have the same members, even if they are referred to differently. In everyday *intensional* (with an *s*) logic, this isn't so. Entities that are equal in extensional logic can't always be interchanged in intensional logic. "Creatures with hearts" and "creatures with kidneys" may refer to the same set of creatures extensionally (i.e., all creatures with hearts may happen to have kidneys and vice versa), but the terms certainly differ in intension or meaning. Likewise, one may promise to arrive in Philadelphia on the day of the wedding, but even though the wedding is President Millard Fillmore's birthday (i.e., they are extensionally the same), it would be an odd person who would describe his date of arrival in this alternative way.

Betwixt the two logics lies a gap we cannot ignore. In mathematical contexts, the number 3 can always be substituted for or interchanged with the square root of 9 or the largest whole number smaller than the constant p without affecting the truth of the statement in which it appears. By contrast, although Lois Lane knows that Superman can fly, and even though Superman equals Clark Kent, she doesn't know that Clark Kent can fly, and the substitution of one for the other cannot be made. Oedipus is attracted to the woman Jocasta, not to the extensionally equivalent person who is his mother. The perspectives of Lois Lane and Oedipus may be limited, but in the impersonal realm of mathematics one's ignorance or, in general, one's attitude toward some entity does not affect the validity of a proof involving it or the allowability of substituting equals for equals.

The logic of history is intensional. Take any historical account of a major event and substitute for incidents and entities in it any extensionally equivalent ones that come to mind. The result will likely be humorous or absurd, such as substituting Millard Fillmore's birthday for any reference to one's wedding day, or substituting the day a local murderer, Craig Rabinowitz, pleaded guilty to killing his wife for any reference to Chinese President

Jiang Zemin's simultaneous appearance in Philadelphia. ("Ah, one of the happiest days of my life was Millard Fillmore's 172nd birthday" or "There were noisy protests at Independence Hall soon after Craig Rabinowitz's guilty plea.") Our view of the event and, in general, the verdict of history depends to an extent on which extensionally equivalent characterization we choose. And which characterization we choose depends on many things, including our psychologies, the historical context, and the history subsequent to the event in question.

Intensional nonsubstitutability also holds for the description of everyday events. When my brother and I were kids visiting our grandparents we would work our way around their neighborhood, taking turns throwing darts at the large trees planted every 25 feet or so along the sidewalks and keeping score of how many we hit. I once convinced him to have this contest in our underwear. He never realized until we'd returned to our grandparents' house that I was wearing swimming trunks under my underwear. In his anger and my gloating, we both accepted that somehow I had appeared less moronic than he during this escapade.

More generally, we all sometimes want, believe, expect, fear, or are embarrassed by something without wanting, believing, expecting, fearing, or being embarrassed by something else to which it is for all (im)practical purposes extensionally equivalent.*

---

*In addition to distinguishing between *intension* and *extension*, logicians and philosophers differentiate between the terms *intension* and *intention* (with a *t*). *Intension* is used loosely to refer to meaning and technically to refer to linguistic contexts in which the substitution of

What is the relevance of all this to probability and statistics? As subdisciplines of pure mathematics, their appropriate logic is the standard extensional logic of proof and demonstration. But for *applications* of probability and statistics—which is what most people mean when they refer to them—the appropriate logic is informal and intensional. The reason is that an event's probability, or rather our judgment of its probability, is almost always affected by its intensional description.

Recall, for example, from the first chapter the choice we have in assigning a likelihood to Waldo's possessing a given characteristic. If we describe him as employee 28–903 in a certain company in country X, 45 percent of whose citizens have a certain characteristic, then it would be reasonable to assume there is a 45 percent probability that Waldo shares this characteristic. But if we describe him as the only person living at a given address and belonging to a certain ethnic group, 80 percent of whose members in the region comprising countries X, Y, and Z

---

extensionally equivalent terms does not preserve truth. The above-mentioned Superman–Clark Kent, mother–Jocasta examples are classic illustrations. The related term *intention* usually is reserved for the characterization of purposive behavior or mental states (desiring, fearing, believing, etc.) directed outward toward something. The terms are closely related; writing of an agent's intentions, for example, establishes an *intensional* (nonsubstitutional) context. He may desire *X* without having the same intention toward *Y*, which is extensionally equivalent. Finally, there is also the everyday English word *intend,* which is one particular kind of intention and thus any discussion of what someone intends in this ordinary sense is also intensional. Herein I will not be overly punctilious about these distinctions (except for that between *extension* and *intension*).

have the characteristic in question, then we probably would conclude the chances are 80 percent that Waldo shares this characteristic. And if we describe him as a specific low-level functionary in a nation-X-wide organization, only 15 percent of whose members have this characteristic, then we likely would state that his chances of having said characteristic are only 15 percent.

These descriptions are extensionally equivalent; they each specify the same individual, Waldo. Which (combination of) inequivalent intensional descriptions we employ and which we take to be most basic is up to us to some extent. Yet choosing such descriptions affects our assignments of probability and everything that flows from them.

For a somewhat different illustration of this point, imagine two statisticians collecting data on when a football team is likely to punt the football. (No knowledge of football is required here.) Being competent practitioners, they come up with seemingly identical results: one statistician concludes that teams punt on *last* down 95 percent of the time, while the other maintains that teams punt on *4th* down 95 percent of the time. (The rules of football specify four downs, so last down and 4th down are the same.) But what if the rules of the game are changed to allow five downs? In this case the first statistician's prediction remains accurate (teams will likely still punt on last down about 95 percent of the time), but the second statistician's finding is incorrect (teams do not punt on 4th down anymore). Although last down and 4th down were extensionally identical under the old rules, their difference in intension or meaning becomes clear when the rules are changed.

What happens when the rules governing, say, welfare are changed and groups of people that were extensionally equivalent are no longer so? Will our carefully collected statistics still be relevant? Or what of economic statistics in a society where breadwinner and husband were once identical, but owing to legal and social changes, are no longer extensionally equivalent?

Statistical results are crucially dependent on intension and context. Uncritical applications of probability and statistics to situations governed by man-made rules and laws can lead to scientistic nonsense. Stories and the logic and rules implicit in them are inseparable from statistics. Specifically, any statistical study on a structured entity—a game, a welfare system, marriages, a historical era—is likely to be fatally flawed if it fails to take the structure into account, say by mindlessly substituting extensionally equivalent entities for one another within the study.

The appropriateness of a mathematical application is continuously vulnerable to criticism and fundamental disagreement; in this sense it is unlike the validity of a mathematical proof but a bit like the varying cogency of a literary interpretation.

INTENSIONAL LOGIC AND COMBINATORIAL EXPLOSIONS

What exactly is intensional logic? Intensional logic is an ill-formed and incompletely understood collection of disciplines that includes certain outgrowths of mathematical logic and philosophy (so-called modal logic, situation semantics, inductive logic, and action theory), parts of linguistics, information theory, cognitive science, psychology, and most important of all, the informal everyday logical intuitions and understandings we all have.

Intensional logic is more tied to context, perspective, and experience than extensional logic and hence requires the use of indexicals: words such as *this, that, you, now, then, here, there,* and last but certainly not least, *I* and *me.* When using intensional logic we must situate the action and the people involved. We must take into account their traits, people and things they know, and circumstances in which they find themselves. Such situating and contextualizing is the analogue of establishing the initial conditions of a scientific law—the height and velocity of a projectile, the temperature and pressure of a gas, etc. However, unlike the case in science where laws are numerous and the initial conditions are often minor details, in intensional logic the contexts, connections, and conditions are much more important than the relatively few "laws" of behavior. The hackneyed refrain "You really had to be there to understand," is often true.

It is not entirely futile to enquire how many such contexts, connections, and conditions exist. (Accepting the short answer "lots and lots" may induce you to skip the next couple of megalonumerical paragraphs.) Some order-of-magnitude estimates hint at the enormousness of this vaguely defined number. If we assume that people can vary along 2 dimensions—have or not have 2 different traits such as shyness and intelligence—then there are 2^2 possible sets of traits someone might have: shy and intelligent, shy and not intelligent, not shy and intelligent, and not shy and not intelligent. If we assume that people can vary along N dimensions—have or not have any of N different traits—then someone might have 2^N possible sets of traits.

If N is as low as 100, 2^{100} is more than a quadrillion quadrillions, and so someone might have more than a quadrillion quadrillion possible sets of traits (not to even mention traits that admit of gradations). And what of the possible connections among people? If there are X people, there are $[X \times (X - 1)]/2$ possible pairs of people, and $2^{[X \times (X-1)]/2}$ possible sets of pairs. Again, if X is as low as 100, there are 4,950 possible pairs of people, and $2^{4,950}$ (a 1 with about 1,500 zeroes after it) possible sets of pairs of people. And with triples of people and more? What about connections or associations among the traits of (collections of) people?

And what about possible situations—the almost unimaginably large number of possible varieties of observing, speaking, buying and selling, making, and so on? Situations are composed of an indeterminate number of different elements, the possible combinations of which are subject to even greater combinatorial explosions.

With such gargantuan numbers about, one might think that conventional statistical approaches would pay handsome dividends, but they don't. From our individual viewpoints our particular constellation of connections almost always seems special, even if from the outside it looks quite ordinary. Knowing ourselves best of all, we are aware of every fine nuance in our backgrounds and circumstances and are usually self-centeredly oblivious to the details of others' backgrounds and circumstances. Although vastly many possible circumstances, connections, and conditions exist, from our individual viewpoints relatively few are sufficiently similar to our own experiences to generate relevant generalizations or statistics. Thus seemingly little statistical guidance is to be gained from them.

Yet we do manage to classify situations, relationships, and people; indeed, we must, else the particularity of any encounter would overwhelm us and render us incapable of carrying on. (Compare the segment on stereotypes in the first chapter.) Informal rules of many instantly recognizable types of situation guide much of our behavior. Somehow we manage to attend to higher-order regularities and bracket off irrelevant details. We often read novels for insight into our own lives from characters who are quite unlike us, superficially. We all understand and use an unformalized, probably not fully formalizable logic that informs our everyday actions and interactions and the narratives that are made of them.

●   ●   ●

Like standard logic, informal logic contains variables. Though students in beginning algebra or standard logic courses sometimes tremble at their introduction, variables aren't any more abstract than the pronouns of intensional logic to which they bear a strong conceptual resemblance. (Likewise, nouns are the analogues of mathematical constants.) Since few people have difficulty with the notions of pronouns or their referents, it would seem that few should have difficulty with variables.

Mathematics, however, has a twist: equational constraints are placed on variables that often enable us to determine their value. Thus, if $5X - 4Y - 3Z + 3(1 + 7X) = 22$

and $Y = 3$ and $Z = 2$, we can find X. (Or, more topically, if
we are told the real-life equivalent of one of a Barbie
doll's four measurements—height, breast, waist, and
hips—we can determine the real-life equivalents of the
other three by measuring and then solving simple algebraic equations. I use the illustration of Barbie because
this elementary fact seems to have been beyond the ken
of reporters whose recent historic news stories dealt with
the keenly anticipated measurements of the new doll.
Contrary to the mystery conjured up in these stories,
whether the manufacturer releases Barbie's new measure-
ment figures or not we at least know their relative magni-
tudes and need stipulate only one to calculate the rest.)

The techniques used to solve these equations and other
more complicated ones have no direct analogue in everyday discourse, although solving mystery whodunits and
making everyday inferences are vaguely similar. Consider
the following: Whoever (X) canceled her (Y's) hotel
reservation knew that she (Y) would be coming for the
celebration, that she (Y) would be arriving late, and that
not having a reservation in her (Y's) name would be annoying to her (Y) and embarrassing to the person (Z)
who invited her. If we know Y and Z, can we discover who
canceled the reservations (X)? Instead of the laws of
arithmetic and standard mathematical logic we use the
more nebulous laws of psychology and intensional logic.

THE BARE BONES OF STANDARD LOGIC

Before further exploring the nature of intensional logic I
would like to sketch the rudiments of standard mathe-

matical logic and, in the process, indicate why it is not suitable for handling everyday situations and stories, contexts, and conversations. First off, standard mathematical logic concerns itself with the meaning of very few words. Each of the astonishing insights that follows is an instance of a mathematical tautology—a statement that is true by virtue of the meaning of the logical connecting words *not, or, and, if . . . then,* and some equivalents: "Either Aristotle had halitosis, or Aristotle did not have halitosis"; "whenever it's not true that either Gottlob or Willard is present, both Gottlob and Willard are absent"; "if it's the case that whenever Thoralf is angry Bertrand is unhappy, then whenever Bertrand is happy Thoralf isn't angry." (The common practice of symbolizing simple statements such as "Aristotle had halitosis" or "Thoralf is angry" with the letters *P* and *Q* inspires the only bit of logic-based public bathroom humor I know: the difference between men and women is that between the statement [*P* and not *Q*] and the statement [*Q* and not *P*].)

Logicians have formalized the checking process by which complex statements—simple statements strung together with the aforementioned connecting words—are judged to be always true (tautologies), always false (contradictions), or sometimes true and sometimes false (contingent). These rules enable one, for example, to mechanically determine what the following advertisement for the Miss Teen America pageant is really saying: "We are searching for a candidate who is either talented and devoted to public service or one who is physically beautiful. Unfortunately, we cannot consider those candidates who are talented and devoted to public service who are not also physically beautiful."

Such rules are of little use, however, for sentences containing relational phrases. The validity of the inference from "All friends of Ludwig Wittgenstein are friends of mine" and "Mae West is Ludwig Wittgenstein's friend" to "Mae West is my friend" does not depend on the meanings of *and, or, not,* and *if . . . then.* Nor do the rules suggest the association between the statements "You can fool all of the people some of the time" and "You can fool some of the people all of the time." These connections are only captured in an expanded logic that encompasses relational phrases involving variables ("X is a friend of Y" or "You can fool X at time Y") and so-called quantifiers (*all* or *some*). In this wider realm the logician Alonzo Church has proved that, unlike the case for tautologies, there can never be a mechanical procedure for determining the validity of sentences or arguments.

As with the connecting words *and, or, not,* and *if . . . then,* complex statements can be built up out of simple ones with relational as well as connecting words, variables, and (symbols for) quantifiers. Consider the misanthropic logical form "X hates Y" defined over the set of individuals and in which each of the variables can be prefaced with a universal *all* or an existential *some.*

If both variables are universally quantified, the form becomes "For all X, for all Y, X hates Y," or, more naturally, the Hobbesian sentiment "Everybody hates everyone." If the first variable is universally quantified and the second existentially quantified, we have "For all X, there is a Y (such that) X hates Y," or the more realistic "Everybody hates someone." Switching the order of the quantifiers in the previous sentence gives us "There is a Y (such that) for all X, X hates Y," or the scapegoat existence proposition "There is someone who is universally hated." If the first variable is existentially and the second universally quantified, the result is the curmudgeonly "There is an X (such that) for all Y, X hates Y," which in slightly better English is "There is somebody who hates everyone." If both variables are existentially quantified, we obtain "There is an X, there is a Y (such that) X hates Y," or the banal "Someone hates someone." What is the colloquial translation of "For all Y, there is an X (such that) X hates Y"?

The addition of terms symbolizing a relation among two or more objects, connecting words and symbols, quantifying words and symbols, and various rules of inference yields a much more powerful logical system within which essentially all present mathematical reasoning may be formalized. This system, whose beauty, elegance, power, and precision I have not even hinted at, is generally termed predicate logic.

And in a usage that has slipped into general employment, statements or discussions *about* the symbols, propositions, or proofs of predicate logic (that statement is false, this proof is valid, etc.) are termed meta-level statements or discussions. Likewise, meta-meta-level statements are about meta-level ones, and meta-meta-meta-level statements are about meta-meta-level ones. A book review is a meta-level discussion of a book, while an article on book reviews is a meta-meta-level discussion of books. Meta-level statements provide a kind of context for whatever is under discussion. The informal term *context* has a somewhat broader, more circumstantial scope, but like the process of moving to a meta-level, the process of placing in a context also can be iterated—a context for the context, and so on. Such processes theoretically can go on indefinitely but in practice have to stop somewhere. (If one of the paparazzi had taken pictures of the other paparazzi taking pictures of the dying Princess Diana, his pictures would have put the others' picture taking in a context and would have been a sort of meta-level picture.)

Unlike predicate logic and mathematics, English is often imprecise, and translation from an English sentence, especially a metaphoric one, to its formalization is often tricky. "All that glitters is not gold" is an example. Does it mean that not everything that glitters is necessarily gold, or that everything that glitters is necessarily not gold? Even the simple English copula *is* may be translated into formal logic in very different ways. Compare the following: "Estragon is Mr. Beckett," where *is* is the *is* of identity—$e = b$; or "Estragon is anxious," where *is* is the *is* of predication—e has the property A; or "Man is anxious," where *is* is the *is* of inclusion—For all X, if X has the property of being a man, then X has the property of being anxious; or "There is an anxious man," where the *is* is existential (in the logical sense)—There is a man and he has the property of being anxious.

The English article *a* is problematic as well, its interpretation sometimes depending on verb tense, for example. The following two arguments are not equivalent, despite having the same form.

> A cat needs water to survive.
> Therefore my cat Puffin needs water to survive.

A dog is barking in the backyard.

Therefore my dog Ginger is barking in the backyard.

Another difficulty associated with translating English
into this artificially restricted formal language derives directly from the austerity of the set of statements that can
be formed out of logical connectives (*and, or, not*), variables, quantifiers (*all, some*), and relational predicates ("X
attacks Y," "W prefers U to V"). Until recently, the situations or contexts in which statements are made have been
neglected. This indifference is in keeping with the timeless and universal nature of mathematical statements and
arguments but hardly is justified when trying to understand human communication and storytelling.

SITUATIONS, SEMANTICS, AND STATISTICS

Moving beyond mathematical predicate logic, a diverse
band of scholars including the philosopher Saul Kripke,
the mathematician Jon Barwise (my thesis advisor at the
University of Wisconsin), and literary theorist Mark
Turner have worked to formalize different aspects of the
context-bound, self-referential, metaphor-laden, agent-
centered, opaquely referential nature of intensional
logic. A goodly number of other scholars could be cited,
but my aims are limited. I merely want to underline the
importance of intensional logic and show that however
ill-formed a subject, it nevertheless grounds our understanding of mathematical applications, specifically proba-

bilistic and statistical ones (and of narratives and informal discourse as well). Such applications are not always as
clear-cut and unproblematic as people generally believe.

Kripke, for example, has developed in *Naming and Ne-cessity* and elsewhere a "possible worlds" theory of meaning that has clarified various notions of necessity (true in
all relevant worlds) and possibility (true in some). This
theory helps elucidate issues involving the naming of entities (Are Mother Teresa and Groucho Marx the same in
all possible worlds?) and the connections among possible
worlds (If the drummer hadn't been high on speed,
would the melee have occurred?). The "possible worlds"
theory has found much resonance in literary theory as
well, where it has been used to shed light on what is
meant by events, circumstances, and characters in a fictional world (Where does Peter Pan live?).

Barwise in his *The Situation in Logic* and others have devised an extension of mathematical logic that builds into
an utterance an explicit reference to its context or situation and permits a situation to be the object of more complicated utterances. In standard logic it is difficult to
capture an elementary assertion such as "Waldo watched
Oscar study" since what Waldo watched is not an entity
such as a person or physical object but the situation in
which Oscar studies. In Barwise's "situation semantics" referring to situations and types of situations is possible and
natural.

The approach also stresses the self-referential aspect of
everyday narrative and conversational situations. "Common ground" or "common knowledge"—the information
well from which each participant in a dialogue draws and

to which each contributes—is a crucial notion. In the usual formulation of this notion, X is an element of information from the common ground occupied by Myrtle and Waldo if Myrtle and Waldo each know X, know that the other knows X, know that the other knows that the other knows X, and so on in a potentially infinite regress. In an alternative formulation X is an element of information from the common ground of Myrtle and Waldo if Myrtle and Waldo each know Y, where Y is equal to the compound statement "X and Myrtle knows Y and Waldo knows Y." Note that Y is defined in terms of itself. Either way, common ground or common knowledge is an inherently self-referential idea that entails more than two people merely knowing the same bit of information and knowing that the other knows it.

However formalized, the situational and self-referential aspects of ordinary human communication (including communication about probabilities and statistics) help to make storytelling and conversation inextricable parts of self-building and culture construction. To communicate with someone it is necessary to empathize with him or her (which of course requires positing the existence of a him or her). One must make reference to the necessary cultural and background knowledge, to the common ground or common knowledge of the participants, and to the particular situation at hand. The understandings involved are delicate and evanescent and the requisite knowledge base Brobdingnagian.

No computer, for example, has ever passed the well-known Turing test, which is often framed in terms of conversations: Picture yourself carrying on a conversation via a television monitor with two interlocutors. Your job is to decide which one's hardware (or physiology) is based on silicon and which on carbon. If you cannot, the computer is said to have passed Turing's test (named after the logician Alan Turing). At least in the foreseeable future, a computer's conversation will quickly reveal its mechanical soul. The amount of tacit knowledge we possess overwhelms our would-be imitators. We know that cats don't grow on trees and that they don't give birth to tractors, that one doesn't put mustard in one's hat or one's socks in the milk container, that toothbrushes are not larger than we are and sold in furniture stores, that even though fur coats are made out of fur and cloth coats out of cloth, raincoats are not made out of rain. All that would be necessary to expose the impostor would be to ask the machine about a few such humanly obvious understandings.

As someone whose Ph.D. was in mathematical logic, I find it fascinating that some practitioners in this field—a bastion of timeless truths and arguments—are formally exploring the notion that communication is a socially mediated process in which context is often critical. Keith Devlin, in his book *Goodbye, Descartes*, has termed this area "soft mathematics," in analogy to the distinction between the so-called hard and soft sciences. Soft mathematics bolsters some of the intuitions long maintained by scholars of literature and the humanities without, however, jettisoning notions of truth and reference.

Such partial rapprochement should not be too surprising; despite common belief, there is little that is inherently antithetical or irreconcilable between literature, storytelling, or conversation and *applications* of logic, mathematics, or statistics. Just as situation semantics attempts to accommodate more of the richness of everyday understanding in an extended formal logical calculus, so a "situation statistics" should be developed that builds in some of the checks on wayward probabilities that commonsense narrative suggests. We wouldn't think much of someone who tried to fit every utterance in a poem or short story into the Procrustean mold of predicate logic. I think a similar opprobrium should attach to those who mindlessly plug facts and figures into statistical formulae and issue unqualified and misleading pronouncements.

A related point concerns how the collection and dissemination of social statistics affects the quantities being measured. Most surveys of religious beliefs subtly discourage the expression of disbelief or unbelief. Sex surveys, to cite another notorious example, are unreliable for the simple reason that people very often lie when questioned about their sex lives by strangers. How else could heterosexual men consistently report more sex partners, on average, than do women? Nevertheless, the publication of such surveys affects people's sex lives and what they are willing to divulge (or invent); it also tends to define their conception of sexuality. Again, the interpretation of statistics is not exempt from the self-referential, incompletely understood rules of intensional logic.

Narrative Common Ground

The Austrian journalist Karl Kraus once remarked, "Psychoanalysis is that mental illness for which it regards itself as therapy." Although I share his low regard for the narrowly scientific merits of Freudianism, the quip does nicely suggest the self-building aspects of intensional logic. Hearing, telling, and ultimately internalizing stories are necessary steps in the construction of a self. We adopt pieces and patterns from others' personas and make them elements of our own, these pieces and patterns having evolved from simple, animal-like propensities. The empathy we feel for our families, our friends, and in an attenuated form, for the larger community makes us human. (Antipathy works the same way, although generally with an opposite gradient—community, friends, family.) Empathy also makes possible human communication and what might be called cognitive coupling, or more prosaically, a meeting of the minds.

As has been indicated, the extensional logic of science is not adequate for describing this cognitive coupling that plays such a big role in storytelling and conversation. In general, we don't just impart information to one another and then draw static inferences from this information about the external world. We "dance" with each other and establish a common ground in which the story or conversation can proceed (or in some cases, not proceed). An example of a conversation's not proceeding very far—but an instance of a dance nevertheless—is the following schematic conversation type.

GEORGE: Hi, Martha.

MARTHA: What's the matter, George? Are you mad at me?

GEORGE: No, of course not.

MARTHA: Yes you are. Why are you mad?

GEORGE: I'm not, I told you.

MARTHA: You are. I can tell by the tone of your voice.
GEORGE: Martha, I am trying not to be angry with you.
MARTHA: See, you're seething with hostility toward me.
Why? What did I do to deserve such anger?
(*George stalks away slamming the door behind him.*)

Not all conversational two-steps involve such entangled antagonism; some are at quite a remove. An acquaintance of mine in graduate school typically would begin what he considered to be a conversation with "Let X be a completely normed Banach space" and then proceed to the statement of a theorem and its proof without making any eye contact whatsoever. During one of these "conversations" the thought occurred to me that I could leave and not be noticed, but I never did. Although it wasn't his intention, this person's "conversations" illustrated the unimportance of conversation and storytelling implicit in standard logic.

A different sort of graduate school story is told by Keith Devlin, and both it and its many variants show how logically subtle our talk tangos can be. Three graduate students are helping their mathematics professor with his gardening. They come in for refreshments and all have smudges of mud on their foreheads. As they sit around the table sipping their drinks, the professor comments that at least one of them has a smudge on his forehead. After several moments they get up simultaneously to wash their foreheads. Since each of them can already see that at least two of them have smudges on their foreheads, how can the professor's comment—which seemingly imparts less information to them than they already have—be informative?

The brief answer is that the professor's comment adds to their *common* knowledge. Specifically, Mortimer, one of three, can meta-reason in the following manner about the reasoning of Waldo and Oscar, the other two students. Mortimer tells himself that if his forehead were clean, then Waldo would see this and reason that if his forehead were also clean, then Oscar, seeing two clean foreheads, would conclude that his forehead was the offending one and therefore immediately get up to wash it. Since Oscar hasn't done so, then my (Waldo's) forehead must be dirty. Oscar can reason in the same way and conclude that his (Oscar's) forehead must be dirty. Since neither Waldo nor Oscar has made a move for the bathroom, Mortimer concludes it is his (Mortimer's) forehead that is dirty.

But the situation is symmetrical, and Waldo and Oscar will go through the same reasoning process to conclude that each of their foreheads is dirty. Assuming that each student is equally astute and reasons at the same pace,* we conclude that they will all get up simultaneously to wash their foreheads. To summarize, Mortimer, Waldo, and Oscar already were aware that at least two of them had smudges on their foreheads, and yet the common knowledge that at least one of them had a smudge was more action-inducing.

The resonant conjoining of the students' thinking makes explicit an important aspect of the gap between stories and statistics and, more generally, of that between literature and science. Pure mathematics and its exten-

---

*The failure of this assumption underlies a bumper sticker I saw recently: "He who laughs last thinks slowest."

sional logic allow for—indeed, even call for—personal detachment, for standing outside a relationship, a governmental policy, a biological phenomenon, an entire galaxy. Mathematics is extricative; it extricates and disentangles us. By contrast, informal intensional logic, the squishy rules of which develop out of life itself, tends to involve us with others, induce us to influence and be influenced by each other, to presuppose both personal sovereignty and a shared social context. Intensional logic is implicative; it implicates and entangles us.

Here is a variant of the smudge story that better illustrates its implicative nature.

A PARABLE OF FURIOUS FEMINISTS AND THE STOCK MARKET

I wrote this parable a week after the precipitous decline in the stock market in October 1997. It takes place in a benightedly sexist village of uncertain location. In this village there are 50 married couples, and each woman knows immediately when another woman's husband has been unfaithful but never when her own has. The strict feminist statutes of the village require that if a woman can *prove* her husband has been unfaithful, she must kill him that very day. Assume also that the women are statute-abiding, intelligent, aware of the intelligence of the other women, and, mercifully, that they never inform other women of their philandering husbands. As it happens, all 50 of the men have been unfaithful, but since no woman can prove her husband has been so, the village proceeds merrily and warily along. Then one morning the tribal matriarch from the far side of the forest comes to visit.

Her honesty is acknowledged by all and her word taken as law. She warns darkly that there is at least one philandering husband in the village. Once this fact, only a minor consequence of what they already know, becomes *common* knowledge, what happens?

The answer is that the matriarch's warning will be followed by 49 peaceful days and then, on the 50th day, by a massive slaughter in which all the women kill their husbands. To see this, assume there is only one unfaithful husband, Mr. A. Everyone except Mrs. A already knows about his infidelity, so when the matriarch makes her announcement only Mrs. A learns something new from it. Being intelligent, she realizes that she would know if any other husband were unfaithful. She thus infers that Mr. A is the philanderer and kills him that very day.

Now assume there are only two unfaithful men, Mr. A and Mr. B. Every woman except Mrs. A and Mrs. B knows about both cases of infidelity, Mrs. A knows only of Mr. B's, and Mrs. B knows only of Mr. A's. Mrs. A thus learns nothing from the matriarch's announcement, but when Mrs. B fails to kill Mr. B the first day, she infers that Mr. A must also be guilty. The same holds for Mrs. B, who infers from the fact that Mrs. A has not killed her husband on the first day that Mr. B is also guilty. The next day Mrs. A and Mrs. B both kill their husbands.

If instead there are exactly three guilty husbands, Mr. A, Mr. B, and Mr. C, then the matriarch's announcement would make no impact the first day, but by a reasoning process similar to the one just described, Mrs. A, Mrs. B, and Mrs. C each would infer from the inaction of the other two on the first two days that their husbands also were guilty and kill them on the third day. By a process of mathematical induction we can conclude that if all 50 husbands were unfaithful, their intelligent wives finally would be able to prove it on the 50th day, the day of the righteous bloodbath.

Now if you replace the warning of the matriarch from the far side of the forest with that provided by the Thai, Malaysian, and other Asian currency problems this past summer, the nervousness and uneasiness of the wives with the nervousness and uneasiness of investors, the wives' contentment as long as their own oxen weren't goring with the investors' contentment as long their own oxes weren't being gored, killing husbands with selling stocks, and the 50-day gap between the warning and the killings with the delay between the East Asian problems and the big sell-off, you have it. More explicitly, the interested financial parties may have suspected that the other Asian economies were vulnerable, but didn't act until someone said so publicly and thereby eventually revealed their own vulnerability. Thus the Malaysian prime minister's speech criticizing Western banks in April 1997 may have functioned as the matriarch's warning and precipitated the very crisis he most feared.

Happily, unlike the husbands in the story, the market is capable of rebirth. Wall Street's subsequent surge suggests that the analogy would be sturdier if the wives could resurrect their husbands after a short stay in purgatory. Such is life, death, buying and selling in the global village.

●　　●　　●

In a similar way, family secrets and political scheming of which all relevant parties are well aware take on a different character when they become public knowledge. So too is this the case with statistical facts (and pseudo-facts), which, of course, are not immune from the change in role they undergo when they become part of the common knowledge of a large enough group of people. With perhaps fewer mental acrobatics than with the mud smudge story, the ferocious feminists, or family and political intrigues, we selectively take in statistics on wealth distributions, health scares, sexual practices, fads, crime incidence, and a myriad of other matters and make them part of the common ground of our various relationships and narratives.

Such has been the case, for example, with public realization of the enormous number of deaths caused by smoking and general acceptance of the widespread prevalence of spousal abuse (which, unlike the situation in the feminist village, generally means wife abuse). Unfortunately, spurious associations, such as that between breast implants and various autoimmune diseases, also may become part of our common ground for a time. The idiosyncratic set of statistics we buy helps us define who we are, however tenuous their connections to reality. Those that we collectively adopt and that thereby become common knowledge motivate us to get up and either wash our faces or kill our mates. The interpretation of probabilities and statistics once again is parasitic on the hazier realms of intensional logic and psychology.

Of course, most of the fineness inherent in stories and conversations isn't statistical but depends in large part on the countless particulars of a situation and distinctive linguistic styles that evolve through interaction with others. Although philosophers warn of the impossibility of a completely private language, semi-private languages are part of the common ground of any two significantly re-lated people and appear in any extended story. How the members of a couple signal their intentions to purchase items and their attitudes toward money would fill a small book. The natural way these understandings arise is illus-trated in the following story.

A young man is on vacation and calls home to speak to his brother.

"How's Oscar the cat?"

"The cat's dead, died this morning."

"That's terrible. You know how attached I was to him. Couldn't you have broken the news more gently?"

"How?"

"You could've said that he's on the roof. Then the next time I called you could have said that you haven't been able to get him down, and gradually like this you could've broken the news."

"Okay, I see. Sorry."

"Anyway, how's Mom?"

"She's on the roof."

FROM AUTOBIOGRAPHY TO
MODELS AND NOVELS

Another area in which private stories and understandings interact with public pronouncements and statistics in murky, self-referential ways is autobiography. Consider the following assumption and question. Suppose I took the trouble to dredge up statistics on early-twentieth-century immigration from Greece, the psychology of firstborns, life in the 1950s and 1960s, the way-beyond-hackneyed baby-boomer phenomenon, parental divorce, and a myriad of other objective issues ostensibly germane to my psychological development. Would supplying these facts and figures to the reader tell him or her more about me than the next few paragraphs of autobiographical narrative? (The following were taken verbatim from an almost-forgotten file of reminiscences on an almost-forgotten relic of mine, a KayPro computer.)

The first-born grandson of immigrant Greeks, I was the focus of their attention and adulation. I was cosseted and courted, and everything in my early years seemed magical and alive and solitary—the pattern of the whorls on my closet door as I fell asleep, the late afternoon sun glinting off the red bricks and rusted black fire escapes of the buildings across the street, the lines on the sidewalk and chips in the curb, the smells of the corner grocery store and of the sewer in the alley, the almost musical silence of everything. I recall the feeling of complete security as my grandmother parted my hair with her fingers and cleaned my cheek with her saliva. I remember too eating tomato salads with my grandfather on the rooftop of the apartment building in Chicago where we lived.

Every night during the air-conditionerless Midwestern summer, friends and relatives would sit out on the tenement steps and talk and I'd sit in the corner, half listening, half daydreaming. In a completely inarticulate way I sensed that most of the adults' concerns were silly, and this mute realization made me feel strangely happy, very safe, and a little superior. My mother was beautiful and my father was a baseball player. Things were very, very good.

•   •   •

Much later: my parents divorced after only 36 years of marriage and four children. It wasn't much of a surprise. They always had been very different, and they became increasingly so, or so runs the litany. My father likes to remark on the passing scene, to write poems, to joke and muse. He has a Pollyannaish Will Rogers sort of outlook on life. My mother is more intense, focused, and preoccupied. She has more chic, more pizzaz. My father, to my mother's chagrin, is not cool. My mother, to my father's, is not warm.

How can I encapsulate their differences? One day when I was about 10 I stayed home sick from school. As soon as everyone had left, my mother turned on the hi-fi and swirled through the house as she listened to Madame Butterfly, the Helen Morgan Story, and a medley of other torch songs of unrequited love. She seemed happy as she danced and talked on the phone and did the housework in a sort of romantic haze. Then my father came home, his suit rumpled, hat askew, tie loosened as always, a characteristic lop-sided grin on his face. "Spahn's pitching tonight. The Braves are going to come out of their slump. You'll see." I love them both.

Whatever one's response to the question about the relative merits of narratives and numbers (mine is this book), it's clear that storytelling and statistics citing afford alternative sorts of insights into alternative sorts of entities, usually individuals versus collections thereof. I've discussed the extensional nature of statistical data, which contrasts with the necessarily intensional descriptions typical of autobiography. The issue of autobiography, however, suggests other more general (dis)similarities between statistical studies and narratives of all sorts, especially novels.

•   •   •

As miniature representations of parts of the world, both statistical models and realistic novels may be more or less descriptive, more or less accurate, more or less suggestive. Of both sorts of simulation (especially of applications involving economic or social issues or in historical novels or romans à clef) we want to ask how much is based on close observation and research and how much on invention and convenience.

Unfortunately, most statistical models and, I believe, most novels are built with off-the-shelf components that are only slightly customized. It doesn't require much more than money to buy some expensive statistical software package, pick some standard model from it, feed in your data, and sally forth with impressive-sounding but nonsensical pronouncements. Devising such software, of course, is an order of magnitude more difficult than using it, but even this is often uninspired. Likewise, attending enough writers' seminars and conferences will enable most aspiring novelists to dress up their most unimaginative stories in some reasonably presentable and fashionable garb.

Frequently as a result, models don't model very well and novels aren't very novel. In both cases what is essential to attaining some insight are telling details. In the model such details are generally structural. Do the assumptions of the model hold "out there" and, conversely, are the salient regularities "out there" being captured by the model? In the novel the details are situational, stylistic, and emotional. Are the actions, thoughts, and conversations depicted plausible, and, conversely, are realistic actions, thoughts, and conversations finding their way naturally into the novel?

Correspondences between the world and its representations are a good thing, but models and novels are most precious when they make predictions or reveal attitudes that are unexpected and counterintuitive. Demographic models that suggest that people with more disposable income are likely to spend more on luxury items, or novels that reveal that married people sometimes have affairs and lie to their spouses about them are not exactly trenchant. More nuances, complications, and stylistic sparkles are needed.

Basic assumptions often can be variously supplemented and interpreted and give rise to different mathematical models and computer simulations. There are a number of different epidemiological models for the spread of HIV infection, for example, each more or less consistent with known facts and common assumptions and each yielding contrary predictions. One model, that of MIT economist Michael Kremer, even makes the not completely implausible claim that in certain cases increased promiscuity among the population at large would slow the spread of infection, presumably because there would be less need for the most sexually active to concentrate their trysts in the most at-risk subpopulations. Similarly disparate predictions result from various econometric models purporting to forecast inflation, unemployment, and other economic variables.

Likewise, almost identical plot lines can give rise to very different novels. Most traditional romances, for example, have the schematic plot: Boy meets girl. Boy loses girl. Boy wins girl. A schematic recipe exists for thrillers as well. Still, the books that result are certainly not all alike. The same is true for answers to the canonical riddle "What's black and white and re(a)d all over?" which number in the hundreds, as M. B. Barrick demonstrated in *The Journal of American Folklore* in his "The Newspaper Riddle Joke." They range from a newspaper to such nonstandard models of the riddle's conditions as an embarrassed zebra, Santa Claus coming down a dirty chimney, a wounded nun, a rabid racist's view of an interracial couple, and a skunk with diaper rash.

The following Sufi story adapted from Masud Farzan's *Another Way of Laughter* humorously illustrates how difficult it is to pin down an indefinitely rich reality with a few assumptions. As in the newspaper riddle, alternative models almost always can be found for any set of assumptions. In philosophy this is sometimes known as the "underdetermination problem."

Since a Roman religious scholar was visiting the court of the Turkic Emperor Timur, the Emperor asked a famous Mulla to prepare for a battle of wits with the scholar. The first thing the Mulla did was to

burden his donkey with books having nonsensical titles. The day of the contest the Mulla appeared in
court with his donkey, and despite the language bar-
rier between them, overwhelmed the Roman
scholar with his native charm and intelligence. The
scholar finally decided to test the Mulla's knowledge
of theoretical matters. He held up one finger. The
Mulla answered with two fingers. The Roman held
up three fingers. The Mulla responded with four.
The scholar waved his open hand to which the
Mulla responded with a closed fist into his palm.
The scholar then opened his briefcase and took out
an egg. The Mulla responded by digging an onion
out of his pocket. The Roman asked, "What's your
evidence?" The Mulla gestured uncomprehendingly
toward his books. The Roman looked and was so impressed at the titles that he conceded defeat.

Since no one had understood any of this, later, after refreshments had been served, the Emperor
leaned over and asked the Roman scholar the
meaning of it all. "He is a brilliant man, this Mulla,"
the Roman explained. "When I held up one finger,
meaning that there was only one God, he held up
two to say that He created heaven and earth. I held
up three fingers, meaning the conception-life-death
cycle of man, to which the Mulla responded by
showing four fingers, indicating that the body is
composed of the four elements earth, air, water, and
fire. Waving my open hand signified that God is
everywhere, and his closed fist into his palm added
that God is also right here among us." "Well then,

what about the egg and the onion?" the Emperor pressed. "The egg was the symbol of the earth (the yolk) surrounded by the heavens. The Mulla produced an onion indicating the layers of heavens about the earth. I asked him to support his claim that the number of layers of heavens is the same as the number of layers of onion skin, and he pointed to all those learned books of which I, alas, am ignorant. Your Mulla is a very learned man indeed." The dejected Roman then departed.

The Emperor, who also spoke the Mulla's language, next asked him about the debate. The Mulla replied, "It was easy, Your Majesty. When he lifted a finger of defiance to me, I held up two, meaning I'd poke both his eyes out. When he held up three fingers indicating, I'm sure, that he'd deliver three kicks, I returned his threat by promising four kicks. His whole palm, of course, meant a slap in my face, to which I would respond with a hard punch. Seeing I was serious, he began to be friendly and offered me an egg so I offered him my onion."

Although story lines and assumption sets containing more numerical information limit their interpretations more than do the fingers and fists in this parable, many partisan political disagreements, for all their statistics thumping, are not any less stark or complete. Sometimes even better than numbers, nonverbal traditions and practices limit the likelihood of misinterpretation. Older religions that rely on custom, tradition, and written scripture, for example, are probably less likely to be egregiously interpreted than newer belief systems that rely on religious writings alone.

Not only are models, novels, and religions possible interpretations of mathematical assumptions, literary plot lines, and holy books, respectively, but at the risk of pushing the obvious I note that there are different levels at which this is true. As the previous story suggests, even within novels there can be alternative models of fictional reality.

Mistaken identity—the little engine that drags countless stories up and down the narrative hill—can be viewed as a way to introduce an unexpected bearer (a nonstandard model) of a collection of attributes, assumptions, and relations and then to set this character out into the book's world. Epistolary novels too, since they consist largely of the letters and diary entries of two correspondents, often contain opposing interpretations (models) of a common set of facts. Samuel Richardson's *Pamela* and *Clarissa,* two of the very first novels in the English language, are of this form. Murder mysteries often depend on radically different interpretations of the same body of facts (or sometimes simply the same body).

In more modern novels the point of view determines to a large extent the interpretation readers are likely to take away with them. Any story of adultery, for example, can be told from at least four natural points of view: the injured spouse, the unfaithful spouse (if they are both unfaithful, from either of the spice), the lover, or an outsider. Imagine *Madame Bovary* from Charles Bovary's point of view. The incompatible interpretations of the marriage are nevertheless often quite compatible with the basic facts of their situation.

The notion of a mathematical model is a useful tool for clarifying the multivalence and surplus resonance of literary works. So is the notion of a meta-level discussion, which, although it goes back at least to the meta-commentary of the chorus in ancient Greek drama, plays an increasingly common role in experimental meta-fiction and even in popular culture (ranging from "You're so vain, I bet you think this song is about you" to David Letterman and Beavis and Butt-head). The standard tools of literary explication—metaphor, metonymy, mood, mode, mimesis, and many more, including alliteration—are considerably more powerful and numerous. Yet there is some evidence that formal work in logic, philosophy, cognitive science, artificial intelligence, and psychology is inching toward some of the (more reasonable) insights of literary theorists—insights that in circular turn are not at all irrelevant to the interpretation of mathematical applications. The border lands are still largely uninhabited, however, even when they are not strewn with enemy land mines.

•　　•　　•

A last point about models and novels: the necessary truths of pure mathematics, statistics, or logic need to be more explicitly distinguished from the tentativeness and uncertainty of other assertions. The most casual observer knows that mathematical statements are either posited (axioms) or proved (theorems), not tested or confirmed in the way that scientific laws and hypotheses are. Statements from the empirical sciences such as physics, history, or psychology are contingent on the actual physical world. At least conceptually, Boyle's gas laws, the fate of the *Titanic,* and the sexual makeup of the American president could easily have been otherwise; not so the propositions that $2^6 = 64$ or that the integral of $(1/\sqrt{2\pi})(e^{-x^2/2})$ between 1 and 2 is .136. Once a mathematical statement is assigned a physical (social, psychological) interpretation in a model (or novel), it ceases to be a necessary mathematical proposition and becomes an uncertain empirical one.

Applications of mathematics, to reiterate, are subject to critique and dispute, but not its theorems. Our failure to appreciate this fact may lead us to a fate similar to that of the now extinct tribe of hunters who, being experts on the theoretical properties of arrows (vectors), would simultaneously aim arrows northward and westward whenever they spotted a bear to the northwest.

The distinction between necessary mathematical truths and uncertain empirical assertions is closely related to another time-honored one in philosophy. An analytic statement is defined as one that is true or not by virtue of the meanings of the words it contains, whereas a synthetic statement is one that is true or not by virtue of the way the world is. "Bachelors are unwed men" is analytic; "Bachelors are lewd men" is synthetic. "UFOs are flying objects that have not yet been identified" is analytic; "UFOs contain small green aliens from a distant solar system" is synthetic. When Molière's pompous doctor explains that the sleeping potion is effective owing to its dormitive virtue, he is making an empty, analytic statement, not a factual, synthetic one. The distinction derives from similar ones drawn by Immanuel Kant and David Hume. Although challenged in modern times by W. V. O. Quine, who has argued that such a distinction is only one of degree and convenience, it still generally marks a profound division between, in Hume's words, "relations of ideas" and "matters of fact."

•   •   •

The next chapter takes a different tack on some of these issues and deals with a few of the implications of information theory, a branch of probability; one is that stories carry information that in an almost literal sense becomes a part of us.

APPENDIX:
HUMOR AND COMPUTATION

(The following is a slight adaptation of a talk I gave at a 1995 conference on computation and humor at the University of Utwente in the Netherlands. I include it here since the affinities among the abovementioned seemingly disparate topics are germane to the connections between stories and statistics, and informal discourse and logic.)

I wrote a little book entitled *Mathematics and Humor*, which was published in 1980 by the University of Chicago Press. In the book I explore the operations, ideas, and structures common to humor and mathematics. Humor is loosely defined as resulting from a perceived incongruity of one sort or another in an appropriate emotional climate,* and mathematics is understood to include logic, mathematics proper, and linguistics. As

---

*In this book I have used the more general terms *situation* or *context* rather than *emotional climate;* have widened the scope from humor to all sorts of narratives; and for the most part have restricted the mathematics involved to probability and statistics, which are of more immediate practical importance and interest to most people than, say, algebraic topology.

far as I know, it is the first mathematical study of the formal properties of humor.

One of the book's themes is that mathematics and humor both provide cerebral enjoyment and can be thought to reside on a continuum of intellectual play. In mathematics, of course, the emphasis is on the intellectual, in humor on the play, and in the middle somewhere are hybrids such as riddles, puzzles, and brainteasers. Both pure mathematics and humor are generally undertaken for their own sake and not for any narrowly utilitarian reasons. Ingenuity and cleverness are hallmarks of both. In different ways, logic, pattern, rules, and structure are essential to both endeavors, as are iteration, self-reference, and nonstandard models. Both use the logical technique of reductio ad absurdum, but in mathematics the focus is on the reductio, while in humor it's more on the absurdum. Long-winded proofs or jokes are generally anathema; economy of expression is prized in both mathematics and humor, as is literal interpretation of terms (responding to a "Lower Your Voice" sign in a library by speaking closer to the floor).

Much of the humor in the book is what Freud would call nontendentious wit, humor in which wrongheaded forms of reasoning, rather than sexual or aggressive impulses, are safely exposed. An example concerns two idiots, one tall, skinny, and bald, the other short and fat, who come out of a tavern. As they start toward home a bird flies over and defecates on the bald man's head. The short man says he's going back to the tavern for toilet paper, whereupon the tall one observes, "No, don't do that. The bird's probably a mile away by now." Another is about a college student who writes his mother that he's taken a speed reading course. She responds with a long chatty letter in the middle of which she remarks, "Now that you've taken that speed reading course, you've probably already finished reading this letter." A third is about the prison guard playing cards with a convict. On discovering that the convict has been cheating, the guard throws him out of jail.

Five years later, in 1985, Columbia University Press published *I Think, Therefore I Laugh.* The book was at least partly intended as a reply to a remark by Wittgenstein that a good and serious work in philosophy could be written which consisted entirely of jokes. If one understood the relevant philosophical point, one got the joke (or parable, story, or puzzle). I declared that humor and analytic philosophy resonate at a deep level (both evince a strong penchant for debunking, for example) and supported this claim with stories and jokes, some exposition on topics ranging from the grue-bleen and raven paradoxes to the distinction between analytic and synthetic statements, as well as the construction of imaginary dialogues between philosophical and comedic luminaries. Humor me as I provide an example of a conversation between Bertrand Russell and Groucho Marx taken from the book.

GROUCHO MEETS RUSSELL

Groucho Marx and Bertrand Russell: What would the great comedian and the famous mathematician-philosopher, both in their own way fascinated by the enigmas of self-reference, say to each other had they met? Assume for the sake of absurdity that they are stuck together on the 13th meta-level of a building deep in the heart of Madhattan.

> GROUCHO: This certainly is an arresting development. How are your sillygisms going to get us out of this predicament, Lord Russell? (*Aside:* Speaking to a Lord up here gives me the shakes. I think I'm in for some higher education.)
>
> RUSSELL: There appears to be some problem with the electrical power. It has happened several times before and each time everything turned out quite all right. If scientific induction is any guide to the future, we shan't have long to wait.

GROUCHO: Induction, schminduction, not to mention horsefeathers.

RUSSELL: You have a good point there, Mr. Marx. As David Hume showed 200 years ago the only warrant for the use of the inductive principle of inference is the inductive principle itself, a clearly circular affair and not really very reassuring.

GROUCHO: Circular affairs are never reassuring. Did I ever tell you about my brother, sister-in-law, and George Fenneman?

RUSSELL: I don't believe you have, though I suspect you may not be referring to the same sort of circle.

GROUCHO: You're right, Lordie. I was talking more about a triangle, and not a cute triangle either. An obtuse, obscene one.

RUSSELL: Well, Mr. Marx, I know something about the latter as well. There was, you may recall, a considerable brouhaha made about my appointment to a chair at the City College of New York around 1940. They objected to my views on sex and free love.

GROUCHO: And for that they wanted to give you the chair?

RUSSELL: The authorities, bowing to intense pressure, withdrew their offer and I did not join the faculty.

GROUCHO: Well, don't worry about it. I certainly wouldn't want to join any organization that would be willing to have me as a member.

RUSSELL: That's a paradox.

GROUCHO: Yeah, Goldberg and Rubin, a pair o' docs up in the Bronx

RUSSELL: I meant my sets paradox.

GROUCHO: Oh, your sex pair o' docs. Masters and Johnson, no doubt. It's odd a great philosopher like you having problems like that.

RUSSELL: I was alluding to the set M of all sets that do not contain themselves as members. If M is a member of it-

self, it shouldn't be. If M isn't a member of itself, it
should be.

GROUCHO: Things are hard all over. Enough of this sleazy
talk though. (*Stops and listens*) Hey, they're tapping a
message on the girders. Some sort of code on the girders, Bertie.

RUSSELL: (*Giggles*) Perhaps we should term it a Gödel
code, Mr. Marx, in honor of the eminent Austrian logician Kurt Gödel.

GROUCHO: Whatever. Be the first contestant to guess the
secret code and win $100.

RUSSELL: I shall try to translate it. (*He listens intently to the
tapping*) It says "This message is . . . This message is—

GROUCHO: Hurry and unlox the Gödels, Bertie boy, and st,
st, stop with the st-st-stuttering. The whole elevator shaft is
beginning to shake. Get me out of this ridiculous column.

RUSSELL: The tapping is causing the girders to resonate.
"This message is . . .

A loud explosion.

The elevator oscillates spasmodically up and down.

RUSSELL: ". . . is false. This message is false." The statement as well as this elevator is ungrounded. If the message is true, then by what it says it must be false. On the
other hand if it's false, then what it says must be true.
I'm afraid that the message has violated the logic bar-
rier.

GROUCHO: Don't he afraid of that. I've been doing it all
my life. It makes for some ups and downs and vice versa,
but as my brother Harpo never tired of not saying: why a
duck?

•   •   •

One conclusion I draw from these and other little vignettes I
might have included is that the attempt to understand, generate, and systematize both the form and content of humor is fully

equivalent to the general problem of artificial intelligence. That is, comprehending humor is in most senses equivalent to comprehending intelligence. Humor permeates our understanding and is indistinguishable from it. I believe Wittgenstein was correct when he wrote that humor should be viewed as an adverb and not as a noun. Even the schematic "definition" of humor— a perceived incongruity in an appropriate emotional climate— makes its universality clear. Few words are more nebulous and encompassing than *incongruity*. Note, however, that the humor project is not an impossible one, only that it's fully as difficult as any in artificial intelligence.

So what should computer scientists do to create a machine that at least metaphorically can be said to have a sense of humor? Let me offer four rather disparate suggestions.

First, they should become less ambitious and work on special-purpose humor recognizers and generators for particular kinds of humor and joke schemas, many of which already can be generated by computer. Elementary combinatorial techniques result in puns (W. C. Fields's recommendation of clubs for children falls out easily), spoonerisms (such as "time wounds all heels"), various random alterations of standard texts (the classic N + 7 game, for example, whereby every other noun in a piece of writing is replaced by the seventh noun following it in some standard dictionary), palindromes and their many dyslexic cousins, perverbs (combinations of two proverbs such as "A rolling stone gets the worm" or "A bird in the hand waits for no man"), Chomskian transformations (Wife: Won't you stop smoking for me? Husband: What makes you think I'm smoking for you?), simple iteration (one minor calamity after another), and also iteration combined with some aspects of self-reference or meta-reference (a tow truck towing a tow truck, a neurotic's worrying about having too many worries, an animal on the endangered species list whose sole diet is a plant on the endangered species list, or an e-mail message whose subject is "See Content" and whose content is "See Subject").

As the units of analysis become more abstract, the problem becomes more difficult. Thus recognizing or generating modal jokes—quips in which the form is at odds with the content (Calm down, goddammit)—would require considerably more programming skill, linguistic sophistication, and background knowledge than would recognizing or generating puns. We're immeasurably far from generating jokes such as the following classic (true) story: A well-known, but here anonymous, philosopher was delivering a talk on linguistics and had just stated that the double negative construction in some natural languages has a positive meaning and in other languages a very negative meaning. He went on to observe, however, that in no natural language does a double positive construction have a negative meaning. To this another well-known philosopher in the rear of the lecture room responded jeeringly, "Yeah, yeah."

This brings me to the second of my four suggestions. I remarked that a sense of economy is generally essential to both mathematics and humor, and one way to measure economy is by employing notions of complexity devised by Greg Chaitin and others (to be discussed further in the next chapter). If two computer programs generate the same sequence of 0s and 1s, for example, the shorter one is generally to be preferred. This is a version of Occam's razor (also to be further discussed), which advises us not to introduce unnecessary entities or complications into our accounts.

The Freudian notion of economy, whereby aggressive or sexual thoughts are succinctly invoked in a disguised way, might with some effort be viewed in this way. Puncturing bombast and pomposity is roughly analogous to finding a pithy remark or a shorter "program" with the same logical content as something longer—which suggests why reductionism and debunking seem to be elements of so much humor. (By the way, the analogue of an incompressibly short program might be a classic epigram that cannot be improved on. Aptly enough, the simple equations that generate the convoluted Mandelbrot fractal have been called the wittiest remarks ever made. Searching for such perfect nuggets might be a good strategy for humor researchers.)

Considerations of economy or brevity also bring to mind those questions on IQ tests and Mensa Society problem lists in which we are given three or four elements in a sequence and asked which of several alternatives is the continuation of the sequence. Since any finite sequence can be continued in any way, every alternative is a continuation. (The fourth term of the sequence 2, 4, 6, . . . might not be 8, for example, but 38, since it could be argued that the Nth term is $[2N + 5(N - 1)(N - 2)(N - 3)]$.) What is required is that continuation which can be most concisely described. Because what constitutes an appropriate language for this description is not usually specified, the question doesn't always admit of a clear-cut answer.

My third suggestion to humor researchers is to pay closer attention to developments in evolutionary psychology. Recent work, for example, has suggested that we have especially keen intuitions about three areas: social cheating, spouse selection, and food. This fact should have relevance for computational humor research. (Incidentally, we should acknowledge that the very idea of "computational humor research" will strike some as humorous and others as menacing.)

1. Consider social cheating. The psychologist Peter Wason has demonstrated that many people are not very good at the following task: Four cards, each with a number on one side and a letter on the other are arrayed on a table before someone. The subject is asked which cards he must turn over to confirm the statement that if a card has a D on one side, it has a 3 on the other. The cards before him are D, F, 3, and 2. Most people turn over the D and 3 cards instead of the D and 2.

Now contrast this task with that of a bouncer at a bar who must throw out underage drinkers. He's confronted with four people: a beer drinker, a cola drinker, a 28-year-old, and a 16-year-old. Which two should he interrogate further? Here it's clear that it is the beer drinker and the 16-year-old who must be interrogated, and jokes in this context (say, involving an idiot bouncer checking on the 28-year-old) obviously are more likely to be understood than ones involving the use of flash cards.

2. How about spouse selection? A man goes to a mathematical geometer and asks for a surface that satisfies the first four axioms of Euclidean geometry and is surprised when the geometer gives him a saddle-shaped surface, which satisfies the four axioms but not the infamous parallel postulate. The once-surprising existence of non-Euclidean models of Euclid's first four axioms can be seen as a sort of mathematical joke.

It is a joke that even Immanuel Kant did not get, but contrast it with the situation of an applicant to a computer dating service who specifies that he wants someone short, gregarious, fond of formal attire, and interested in winter sports. He is surprised, of course, when the computer offers him the name of a penguin. Again, jokes in this context are easier to grasp than others with the same formal structure.

3. Then there's food. Recall that Bertrand Russell once termed the "scandal of philosophy" David Hume's insight that the standard justification of scientific induction is itself inductive. We expect the future to be like the past in certain ways only because past futures have been like past pasts in these specific ways. Contrast this with the more nutritive situation of the woman who went to the doctor pleading for help because her husband thought he was a chicken. When the doctor asked how long he'd thought this, the woman replied that it was for as long as she could remember. "But why didn't you come to me sooner?" he asked, to which she replied that she would have, but she needed the eggs (just as we accept induction without a convincing rationale because we need it).

A joke from Leo Rosten's *The Joys of Yiddish* comes close to combining food, spouse selection, and social cheating. A young man asked a rabbi for help in holding conversations with young women, and the rabbi said that the best topics were food, family, and philosophy. So the young man calls on a young woman and blurts out, "Do you like noodles?" No, she says. He follows with, "Do you have a brother?" No, she says. Then he asks, "If you had a brother, would he like noodles?"

One more example to demonstrate my thesis that any account of humor will also be an account of cognition: In his book *Full House*, Stephen Jay Gould writes of the misconceptions caused by focusing only on averages or extremes of a distribution. Doing so often imparts a spurious sense of movement, of decline or growth, that is not borne out by looking at the statistical distribution as a whole. Consider, for example, his explanation for the disappearance of the .400 hitter in baseball in which he argues convincingly that the absence of such hitters in recent decades is not due to any decline in baseball ability but rather to a decrease in the disparity between the worst and best players (both pitchers and hitters). When almost all players are as athletically gifted and well trained as they are today, the distribution of batting averages shows less variability than it did in the past, and hence .400 averages are rare. (The average of all batting averages has remained relatively constant over the decades, sometimes through fiddlings with the rules.)

This last is a somewhat abstract point, but a joke statistic cited earlier depends on the same understanding and is much easier to apprehend. The statistic is that the average resident of Miami, Florida, is born Hispanic and dies Jewish. Here the suggestion of spurious movement (conversion, in this case) is more easily resisted, but the intellectual content is the same.

My fourth and final suggestion concerns catastrophe theory representations of humor. An interesting topological theory discovered by the French mathematician René Thom in 1975, catastrophe theory concerns itself with the geometric description and classification of discontinuities (such as jumps, switches, reversals, e.g.). As I demonstrated in *Mathematics and Humor*, catastrophe theory provides a sort of mathematical metaphor for the structure of (some) jokes.

The reason for this theory's usefulness is not hard to find. Running through the logic of humor is the idea of an abrupt switch or reversal of interpretation resulting in the sudden perception of some situation, statement, or person in a different and incongruous way. This interpretation switch may be accompanied by the overcoming of a mild fear, as when one realizes that what seemed threatening is not truly so; or the conversion may come upon solving a riddle. Sometimes a release of feelings accompanies the switch, especially when making or laughing at an aggressive or sexually offensive witticism. The interpretation reversal may signal the expression of a playful approach to a situation. At other times the achieving of self-satisfaction is a concomitant of the reversal, as in Hobbes's "sudden glory" resulting from someone else's (mild) misfortune. In all these cases a sudden interpretation switch brings about a release of emotional energy that usually takes the form of laughter or sometimes groans.

Thom's primary result, his classification theorem, describes what can happen when some quantity is discontinuous, satisfies certain mild constraints, and depends on no more than four factors. The seven geometrical figures that the theorem shows exhaust all possibilities are of limited practical use since they are qualitative and difficult to quantify. At first glance each simply looks like several surfaces layered and intersecting in various ways. Nevertheless, in a loose sense, they give us the shape of certain jokes. The different possible meanings of a simple joke's preamble, for example, can be measured along the x and y axes, and its interpretation by the listener can be measured along the z axis. (Alternatively, the z axis can be taken to be a measure of neurological excitement.) The interpretation switch occurs when, at the joke's punch line, the interpretation path "falls off" the top surface of the geometrical figure to the bottom one. The ambiguous set-up of a simple joke often occupies a cusp-shaped region allowing for two possible interpretations to develop. If the joke is not told correctly (e.g., if

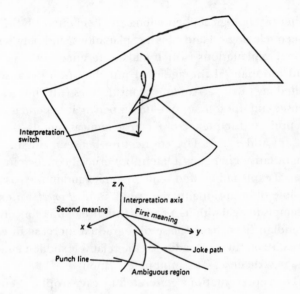

elements of it are related out of order), then the switch from the ostensible interpretation to the hidden one does not occur and the joke "falls flat." This colloquial phrase is one of several metaphors that can be given a mathematical meaning. Light is also cast on issues of comedic timing and of "being on the edge."

The cusp diagrams of catastrophe theory also reflect the noncommutativity of most communications. The extensional logic of science is commutative; the order in which statements are presented does not affect the likelihood of their truth. This is not so in everyday intensional logic, where "first impressions" often change the hearer's mind-set. In the sequence *skyscraper, cathedral, temple, prayer,* for example, the word *prayer* is most likely to be seen as not belonging, whereas in the sequence *prayer, temple, cathedral, skyscraper,* the word *skyscraper* is most likely to be seen as not belonging.

These representations may provide an important link between qualitative and quantitative accounts of humor and

other psychological phenomena and help diminish the gap between teleological and causal explanations, and between intentional explanations (which make reference to the intentions
and rationale of the agent—I punched him because he in-
sulted my wife . . .) and mechanistic ones (which refer only to
causes and not to reasons—the upper right appendage, after a
complicated sequence of electrochemical signals, traveled in
an arc and . . .). The connection between qualitative and
quantitative change is a latent but crucial concern in psychological explanations and some therapies and in the nascent discipline of intensional logic. Freud, in his *Project*, for example,
attempts to deal with it. A typical instance of this concern with a
Freudian flavor is the case of a gradual increase in external
pressure on the ego leading to the relatively sudden emergence
of a new defense, say symptom formation.

The representations generated by catastrophe theory have
been rightly criticized as being unfalsifiable and terminally
hazy. (As such they fit in quite naturally with Freudian psychology!) Nevertheless, I do think that they, and variants and refinements of them, warrant further study. The beginnings of
any such theory are likely to be halting and somewhat stupid,
but the idea of a geometric representation of thoughts and
jokes is too attractive to relinquish easily.

I close this little survey with the humor of absurdity, which
arises whenever the disparity between our pretensions and reality is too stark for even the most blinkered of us to ignore. If
eventually there are computers with sufficient intelligence,
complexity, and experience to develop a sense of humor,
they'll also suffer (or possibly exult in) this sense of absurdity.
But since recognizing and generating humor is more or less
equivalent to thinking and communicating, it'll be a long while
before we see a sillycom (silicon comedian). It'll be even
longer before we see it joking ruefully in the face of its own imminent demise, say because of the beta-testing of Version
732.116.jgd.iv.

4
βetweeŋ meaŋiŋg aŋd iŋformatioŋ

Of course, the entire effort is to put oneself
Outside the ordinary range
Of what are called statistics.
—Stephen Spender

THE VERTIGINOUS CHASM between a Chekhov short story and a sequence of 0s and 1s is not spanned but is clarified a bit by ideas common in information theory. An amalgam of probability and computer science, information theory illuminates a few of the connections among stories, statistics, and selves by providing a kind of skeletal X ray of these fleshier relationships.

Just how skeletal and how fleshy can be illustrated by pointing out how naturally (or perhaps unnaturally) a single sequence of 0s and 1s can encode an arbitrarily large body of text. Consider that a typical Chekhov short story, for example, probably contains around 25,000 symbols: upper- and lowercase letters, digits, blank spaces, and punctuation marks. Each of these symbols can be represented by a sequence of 0s and 1s of length 8 (P has code 01010000; V, 01010110; b, 01100010; ", 00100010; t, 01110100; &, 00100110, etc.). Thus, if we simply concatenate all these sequences, we come up with an approximately 200,000-symbol sequence of 0s and 1s that we can take to be the representation of the story.

Or we can be much more ambitious and arrange all the books in the Library of Congress alphabetically by author and publication date and then concatenate their sequences to come up with a colossal sequence representing all the information in the Library of Congress. Since any sequence of 0s and 1s can be thought of as a single binary number, all the information in the Library of Congress is encoded by this number.

Information theory provides considerably more useful insights regarding the information content of stories, the limited complexity of brains, the wisdom of not looking *too* hard for pattern, and the strange notion of order for free. Before getting to these, however, I suggest that the reader keep in mind in what follows the relation between (an informal sense of) information and (our notion of) self.

One preliminary and indirect approach to this relation between information and self is via false memory syndrome. The ease with which false memories can be implanted in suggestible people by therapists (and others) has been much in the news recently. Yet we are all suggestible to some degree, and I suspect that movies, magazines, and television can do to us what well-meaning therapists sometimes do to their patients. Exposed again and again to so many scenes so vividly and memorably presented, we adopt some of them as our own.

Memories that are integrated and belong to one individual are being replaced by informational shards that belong to no one and everyone. One of mankind's most important inventions is the idea of individual selves who possess individual memories and stories. If individual memories are replaced to any great extent with free-floating, often celebrity-saturated memories in a fragmented world comprised of individuals with little self-definition, then something precious—our individual selves and stories—is diminished, while something of dubious value—our collective television- and movie-mediated selves and stories—is enhanced.

Of course, the whole concept of self may be passé in this age of the planetary webified megalopolis. What is a self in this simultaneously communal yet fractured age? Isn't Steven Wright's quip about his plans to write an unauthorized autobiography emblematic? Many theorists, such as computer scientist Marvin Minsky, philosopher Daniel Dennett, and cognitive psychologist Steven Pinker, argue more or less persuasively that the self is "simply" an assembly or congress of small, semi-sovereign processes whose blind clashes and hagglings result, via some strange, poorly understood sort of deliberation, in an idiosyncratic whole. The Self as the U.S. Senate, a parliament, a diet, a Knesset! We are the laws we adopt. We are the stories we hear. We are the information we process.

I can't remember now where I was going with this, but I do recall the time I was with Madonna on an ornate balcony in Buenos Aires, where she was secretly directing the FBI's activities in Waco, Texas.

Information:
A Lecturer's Insights,
a Girl's Restlessness

What exactly is information? A less ambitious question is, Which is more informative, an overheard snatch of a bar patron's (let's call him Cliff) somewhat inebriated soliloquy, or a single randomly generated image on a TV screen? Assume Cliff has a vocabulary of 20,000 words, and that he utters 1,000 of them. Further assume a TV screen has 400 rows and 600 columns of pixels, each of which takes one of 16 brightness levels. So, again, which has more information content—the 1,000 words or the single image on the TV screen?

The answer, according to the usual definition of information content (due to Claude Shannon, appropriately enough a communications engineer at Bell Labs in the 1940s), is that Cliff's words contain at most 14,288 bits of information (if he utters them at random), whereas the TV image contains up to 960,000 bits. I will omit the probabilistic definition of information content and merely note that it depends on the number of possible states a system—Cliff's speech or the TV image, in this case—can be in and the probability of these states. If a simple message can be in a Yes state or a No state, each with probability 1/2, its information content is 1 bit. More generally, the information content of a message is the number of Yes–No questions that need to be asked to determine what the message is.

I thought of similar examples and calculations recently while sitting in the audience at a public lecture by a renowned economist. The speaker was making a perspicacious point about the falsity of some piece of conventional economic wisdom while a little girl seated with her parents in front of me was haphazardly flexing her index finger, playing with strands of her hair, doodling in a coloring book, and looking idly around the room. There was so much more information content in her desultory actions than in the speaker's utterances that I lost track of the argument.

And yet one rebels at the thought that the little girl's fidgeting (unintentionally engaging as it was), the image on the TV screen, or the listings in a book of random numbers contain so much more information than the lecturer's thesis, Cliff's alcohol-induced pronouncements, or Chekhov's "The Lady With the Dog." The gap between our intuitions of what is meaningful and the mathematics of information content has two sources.

First, there is a growing number of different mathematical definitions of information, ranging from Shannon's notion of information content as a measure of uncertainty or surprise to Gregory Chaitin's definition of complexity, which will be further discussed. These notions are useful and appropriate in different contexts (and by the lights of some, it is certain that the girl's restlessness did have more information content than the economist's lecture).

A second, more fundamental problem derives from the units that we take to be atomic in different realms of endeavor. In formal contexts the units of discourse are bits, program steps, and iotas of time or distance, whereas in everyday life they are basic actions, elements of a scenario, and simple story parts. Unlike the latter higher-order units, neither the arbitrary array of gray pixels on a TV screen, nor the minutiae of the little girl's aimless gestures, nor the sequences of more or less random digits in a phone book *mean* much to us in and of themselves. Without a natural rootedness in some recognizable human context, they mean even less to us than the numbers cited by George Carlin's imaginary sportscaster: "Folks, Nick here. Running out of time, so I'll have to rush with the scores—4 to 2, 6 to 3, in a big blowout, 15 to 3, 8 to 5, 7 to 4, 9 to 5, 6 to 2, and finally in a real cliff-hanger, 2 to 1. And this just in, a partial score: 6."

Can we somehow make narrative units—actions, scenarios, very simple stories and plots—serve as "atoms" in some higher-order analogue of information theory? Although more inherently meaningful than bits and numbers, narrative units still would require some context and meaningful relation to us. While not without problems, it is not necessarily impossible. Something like this relation obtains between computer languages such as Pascal or C++ and a machine's code language of 0s and 1s. Just as these higher-order languages contain terms for common sorts of arithmetic processes, "story languages" would have terms for basic actions such as moving, hitting, taking, going inside or outside, and so on.

In *The Literary Mind* literary theorist Mark Turner discusses some of these basic actions and plot elements. He shows that the parablelike correspondences between them and more complicated stories are essential not only to literary efforts but also to everyday thought and the creation of meaning. Our understanding, Turner argues, derives from the projection of atomic and blended stories onto new situations, much as a proverb such as "When the cat's away, the mice will play" acquires meaning when uttered, say, in a classroom with a substitute teacher. Perhaps when intensional logic is better understood, some of the same information-theoretic notions so useful in technical contexts will be as fruitful in understanding and analyzing stories, statistics, and minds.

• • •

Even as we grant that there is vast scope for formalizing storytelling and identifying higher-order atomic units, we should also grant that more formalization will be dependent on the natural, tentative way in which we decode narrative accounts of events. In much the same manner that probabilistic and statistical notions were distilled out of prior colloquial conceptions or that standard mathematical logic grew out of informal argumentation and conversation, first are the glimmerings of intuition, then refinement of the resulting ideas, and finally, a formal system with clear-cut rules and categories.

The computer scientist David Gelernter has written that the study of the Talmud with its layered, ever more nuanced and cross-indexed stories, parables, conundrums, and commentaries, provides as good a preparation as any for rigorous scientific and mathematical reasoning. The observation can be generalized in two ways. The same is as likely to be true, I submit, of any sufficiently rich text that is pored over with sufficient intensity and thoroughness. Furthermore, the study of such texts is good preparation not only for scientific and mathematical reasoning but also for the construction of a more formal, information-theoretic account of story-telling.

Some might say that we should do away with these open-ended hermeneutic studies; just bring on the shiny formal tools (logical, statistical, informational, whatever) and get on with it. But—I repeat—whatever they turn out to be, the clean information-processing aspects of under-standing stories will depend on the murky interpretive ones. We cannot simply concoct an arbitrary formal the-ory of stories and apply it haphazardly. For similar rea-sons, a statistical software package is dangerous in the hands of someone who has not developed a feel for the various statistical measures the package helps to calculate or hasn't a clue about the relevant variables, populations, or social structures. Note that the terms *feel, relevant,* and *social structures* are informal intensional ones.

I recently ran into a former student of mine outside a campus computer lab. He had received a generous D+ from me in a probability class and was making some ab-surd claims to his friend about the data he was analyzing and throwing around technical terms in a hopelessly con-fused yet superficially impressive way. Similar caveats can be made about narrow and misleading interpretations of information theory.

The gap between meaning and information (of all sorts) is closed somewhat by conceiving of the latter as a distillation of the former, a distillation that needs the soil and water of context to grow. We must understand a situ-ation or situation type thoroughly before we go on to construct a useful formal system in which, say, the information content of a story can be roughly and reasonably quantified. We want to avoid the conceptual equivalent of using a buzz saw to rip a piece of paper or, worse, sew a button on a shirt.

In any case, the inexhaustible source of information is the unmediated world out there. By intelligently reducing parts of it to formal calculi and systems, we tame larger and larger tracts. Still, even when we carry out the reduction thoughtfully, we bring order only to our trim hedgerow in the celestial landscape. Our cognitive homes generally are as unnaturally neat and comfortable as our physical ones.

Trying to acknowledge some of the disorder outside our doors and domesticate it is the task not only of science but also of art and literature; it can even be seen to have been the accomplishment of modernism.* Seurat's pointillism, Schöenberg's atonalism, Kandinsky's abstractions, and Joyce's stream-of-consciousness are just a few examples of this expansion of our psychic dwellings to encompass more of the chaos outdoors. Although they came a little later, so too are Shannon's bits of information.

---

*It is interesting to contrast the inclusiveness of modernist aesthetics with narrower classical prescriptions, in particular with Aristotle's famous ideas on plot presented in the *Poetics,* in which he states that if a plot is well-constructed, any displacement, irrelevancy, substitution, or omission of an incident will destroy the work's unity. These have no place in a story. Similar Aristotelian pronouncements can be made for other arts as well. Our acceptance of extraneous and superfluous bits of information has grown considerably with modernism (electronic or Cagelike "melodies," e.g.), but we still have a healthy intolerance for substitution and irrelevancy in everyday explanations, most story-telling, and other ordinary intensional contexts.

CRYPTOGRAPHY AND NARRATIVES

Information theory suggests the closely related field of codes and cryptography, which has interested a surprising number of authors. Edgar Allan Poe devoted some of his journalistic work to decrypting codes, and they played a key role in "The Gold Bug" and other stories. Jules Verne, William Makepeace Thackeray, and Sir Arthur Conan Doyle also wrote of codes in several of their works, as have many modern authors, including Richard Powers in *The Goldbug Variations.* Since codes, like coincidences and magic tricks, simultaneously appeal to our rational, analytic natures and to our sense of the spooky and mystical, many of the aforementioned observations about the insignificance of most coincidences apply, of course, to codes as well. There is a difference, for example, between my noticing that Poe and I share the same middle name (albeit with different spellings), that he has some connection to my first name as well since he was adopted by a John Allan, that his last name and mine both begin with P, that he lived and I continue to live in Philadelphia, and my ascribing any significance to these coincidences.

My concern with cryptography here, however, is quite different; it is to see what results when we view cryptography as a sort of analogue to literary criticism. That is, assume that encrypting a message corresponds to writing a story with a nonobvious meaning, and decrypting a message corresponds to unlocking the secret of the story. (I do not claim that either endeavor can be reduced to the other.)

One of the simplest and easiest-to-decrypt codes is a so-called linear substitution, in which each letter is replaced by another, let's say, 9 letters ahead of it in the alphabet. For E, F, and G, for example, we'd substitute N, O, and P, respectively. For letters near the end of the alphabet we would "wrap around" to the beginning, substituting B, C, and D for U, V, and W. Given stories' higher order of abstraction, the irreducible vagueness of reference of their units of information, and their yet-to-be developed logic, the closest analogue of a linear substitution code might be a quite literal roman à clef with simple substitutions for characters, places, and times, or perhaps a very rigid allegory or parable.

At the other extreme of decryption difficulty is what's called a one-time pad, in which a message N letters long is encrypted by substituting, for each letter in the message, a different randomly selected letter from 1 to 26 letters ahead of it in the alphabet. The secret code is thus itself N letters long. Assume, for example, the message is the 46-letter line "Either forget that damn lawsuit or it is over between us," and the key is a sequence of 46 random numbers between 1 and 26, say 2, 3, 8, 19, 1, 23, 5 . . . , indicating the number of letters to skip ahead for each letter in the unencrypted message. Then the encrypted message would be "Glbafo k . . . ," since G is E plus 2, l is i plus 3, b is t plus 8, a is h plus 19, and so on. If we assign a symbol for the spaces between letters, the one-time pad is a virtually unbreakable code no matter what the length of the message. Its only defect is that the code must be as long as the message to be encrypted and the key must somehow be communicated to the decrypter.

Again, the story analogue takes account of the abstraction, nebulous reference, and alternative narrative logic of stories. Nevertheless, this method of encryption with a one-time pad is essentially not an encryption at all, since any two messages with the same number of letters could be considered "encryptions" of each other, just as any two stories with the same number of narrative elements could be considered encryptions of one another. Only a demented conspiracy theorist can read a story on new advances in treating diabetes and claim it is in fact a story about the Trilateral Commission's effort to undermine the Aryan Brotherhood.

Somewhere between an unbreakable but impractical one-time pad and an easily decrypted but occasionally still useful linear substitution are more modern codes that depend on what are known as trapdoor functions: they swing easily in one direction but not in the other. Multiplication of two very large prime numbers is an example, since finding their product is easy if we are given the two numbers, whereas finding the numbers given only their product is difficult. Checking to see whether any proposed factorization of the products truly works (as opposed to actually finding one) is also easy. Products of primes are used by banks, businesses, and the military to encrypt material, whereas the prime factors are needed to decrypt it. The analogue of breaking a code of this sort is more like good literary criticism or biographical sleuthing (as in Nabokov's *Pnin*, e.g., which mirrors, transforms, and exalts a part of the author's life). Like factoring a product, insightful criticism is difficult to do, but like checking a proposed factorization, it is reasonably easy to recognize.

I suspect that the more integrated and richly developed a narrative is, the more resistant it is to decoding. (The "prime numbers" required to do so are very long.) The story does not admit of straightforward substitutions or of the decontextualization and atomization that make decoding easier; the same words or phrases mean different things as the story develops. If a few sentences are missing from novels by John Updike and Tom Clancy, for example, more guesses generally will be required to fill in the former omission than the latter.

One final analogy between numerical decrypting and literary exegesis also can be phrased in terms of guesses and questions. In the game of Twenty Questions one player chooses a number between one and a million and the other attempts to determine what the number is by successively asking Yes–No questions of the form, Is the number more than N? Twenty questions are always sufficient, since 2^{20} is slightly greater than a million. More fully: one question is sufficient to reduce the number of possibilities to 500,000, two questions to 250,000, and so on, until twenty questions reduce the possibilities to a single number.

So far, so easy. What happens, however, if the player who chooses the number is allowed to lie once or twice or even more frequently? Even with the correct strategy (which is far from obvious) the number of questions the guesser needs to determine the chosen number grows rapidly, and it is not a trivial problem to calculate what this number is. More complications arise if the questions and lies are allowed to be more complex. One moral this mathematical nugget may have for novelists (the analogues of number choosers and encrypters) is that they should employ unreliable narrators sparingly. Unless the stories are intended to be terminally opaque and ambiguous or instances of "magical realism," a few "lies" by a protagonist/narrator usually are enough to create an agreeable amount of doubt and indeterminacy in the minds of readers (the number guessers and decrypters).

Yet, as we shall see, even when all narrators are perfectly reliable one must resist the tendency to put excessive effort into decoding a story or discovering hidden meanings in it.

OF OCCAM AND TWO-BIT STORIES

How would you describe the following sequences to an acquaintance who couldn't see them?

(1) 0 0 1 0 0 1 0 0 1 0 0 1 0 0 1 0 0 1 0 0 1 0 0 1 0 . . .
(2) 0 1 0 1 1 0 1 1 0 1 1 0 1 1 0 1 0 1 1 0 1 1 0 1 0 . . .
(3) 1 0 0 0 1 0 1 1 0 1 1 0 1 1 0 0 0 1 0 1 0 1 1 0 0 . . .

Clearly the first sequence is the simplest, being merely a repetition of two 0s and a 1. The second sequence has some regularity to it—a single 0 alternating sometimes with a 1, sometimes with two 1s; whereas the third sequence is the most difficult to describe, since it doesn't seem to evince any pattern at all. Observe that the precise meaning of [. . .] in the first sequence is clear; it is less so in the second sequence, and not at all clear in the third. Despite this, let's assume that these sequences are each a billion bits long (a bit is a 0 or a 1) and continue on "in the same way."

Heedful of these examples, we can follow the Russian mathematician A. N. Kolmogorov, of probability fame, and the computer scientist Gregory Chaitin, the author of *The Limits of Mathematics*, and define the complexity of a sequence of 0s and 1s to be the length of the shortest computer program that will generate (i.e., print out) the sequence in question. (The phrase *algorithmic information content* is often used instead of the term *complexity*.)

All of which means that a program that prints out the first sequence can consist simply of the following short recipe: print two 0s, then a 1, and repeat 1/3 of a billion times. Such a program is quite short, especially compared to the length of the billion-bit sequence it generates. The sequence has very low complexity.

A program that generates the second sequence would be a translation of the following: print a 0 followed either by a single 1 or two 1s, the pattern of the intervening 1s being one, two, two, two, two, one, two, and so on. If this pattern persists, any program that prints out the sequence would have to be quite long so as to fully specify the very complicated "and so on" pattern of the intervening 1s. Nevertheless, due to the regular alternation of 0s and 1s, the shortest such program will be considerably shorter than the billion-bit sequence it generates, and so the sequence has higher complexity than the first, but not the highest possible for its length.

The third sequence (by far the commonest sort) is different. This sequence, let us assume, remains so disorderly throughout its billion-bit length that no program we might use to generate it would be any shorter than the sequence itself. All any program can do in this case is to dumbly list the bits in the sequence: print 1, then 0, then 0, then 0, then 1, then 0, then 1, and so on. A sequence like the third one, which requires a program as long as itself to be generated, is said to be random and has the highest possible complexity for its length. (The one-time pad in the previous example where the code is as long as the message to be encrypted is random.)

In some ways sequences like the second are the most interesting, since, like living things, they display elements of order and randomness. Their complexity (the length of the shortest program that generates them) is less than their length, not so small as to be completely ordered nor so large as to be random. The first sequence may perhaps be compared to a diamond or a salt crystal in its regularity; whereas the third is comparable to a cloud of gas molecules or a succession of coin flips in its randomness. The analogue of the second might be a lily or a cockroach, which manifest both order and randomness among their parts.

•   •   •

These comparisons between sequences of various types and other entities are more than what some might term mere metaphors. (*Mere* probably is not the word to apply to something that can be as overwhelmingly powerful as a metaphor.) Gradually we've come to the now commonly held belief that everything can be reduced to information, 0s and 1s, bits and bytes, not atoms and molecules. The introduction of an Arabic numeral for zero, of symbols for rests or silences in musical notation, of empty space and vacancy in later medieval painting, and of Leibniz's metaphysical notion of nonbeing are different instances of this same realization.

Most phenomena can be usefully described via some code, and any such code, whether the molecular language of amino acids, the letters of the English alphabet, or the elements of a yet to be defined "story language," can be digitalized and reduced to sequences of 0s and 1s. Proteins, coded messages, and mystery novels, expressed in their respective codes, are sequences akin to the second one in the example, evincing order and redundancy as well as complexity and disorder. Similarly, complex melodies lie between simple repetitive beats and formless static. (Even programs that generate sequences of 0s and 1s can themselves be encoded as sequences of 0s and 1s.)

One might even conceive of the whole of science in this way. Ray Solomonoff, Chaitin, and others have theorized that a scientist's observations could be encoded into a sequence of 0s and 1s according to some protocol. The goal of science then would be to find good theories (short programs) capable of predicting (generating) these observations (sequences). Each such program, so the conceit goes, would be a scientific theory and the shorter it was, relative to the observational phenomena it predicted, the more powerful a theory it would be. This, of course, is a restatement of Occam's razor, the principle that unnecessary entities and complications are to be eliminated.

Random events would not be explainable or predictable, except in a very Pickwickian sense, by a program that simply listed them. Note that in effect this is what happens when people advance impossibly convoluted theories to account for some essentially random observations. Presented with an oversized checkerboard with squares randomly colored red or black, for example, some people will concoct a Ptolemaic explication far longer than a simple listing of the red and black pattern. They are struggling too hard to account for an observed phenomenon if the account they come up with is essentially as complex as the phenomenon itself.

Also, if the phenomenon in question is not random (and most phenomena of interest are not), there is even less reason for long, unnecessarily elaborate explanations or predictions of it. A mnemonic, like a theory, should not take longer to memorize than that for which it is being used.

The principle of logical parsimony has some relevance for literary analyses and deconstructions as well. I readily grant that in a short story, say, much more appeals to us than its information content or complexity. Nevertheless, exegeses that are considerably longer than the story itself begin to run afoul of the complexity-theoretic ideas just described. It is no trick at all to generate a sequence with a program whose complexity is much greater than that of the sequence. The voluminous criticism that sometimes attaches to a story or the torrential media commentary that surrounds some events therefore must say more about critics and commentators or about other matters than it does about the story or events in question. It can also be mind-numbingly repetitive.

In the last chapter I wrote of the indefinitely many continuations of a numerical sequence that are possible if the rules used to generate the continuations are allowed to be sufficiently long. The simple sequence 2, 4, 6. . . , for example, can be taken as presaging the coming of the millennium, since one can insist that 2,000 is the next number in it. The first four terms (i.e., where $N = 1, 2, 3, 4$) in the sequence given by the complicated rule $[332 (N-1)(N-2)(N-3) + 2N]$ are 2, 4, 6, 2,000. Given an appropriate language within which to have them, the same restrictions on length and complexity apply to discussions of literature. A 400-page explication of a 25-page story is sufficiently long to draw almost any conclusion one might desire from the story. More likely, the bulk of the explication is about the author's sensibilities, or an attempt to relate the story to other works or issues, or is merely, I repeat, repetition.

OUR COMPLEXITY HORIZONS

If the scope of the response to a story is broadened to include details of the author's life, related works and influences, and historical, aesthetic, and political issues, then the number and lengths of the works on a particular story are less unreasonable. A rich story will always elicit commentary, since, as Boris Pasternak put it, "What is laid down, ordered, factual, is never enough to embrace the whole truth." Nevertheless, it should be recognized that a story's information content and its complexity are limited, and the source of the commentary it elicits and the open questions it invites exists outside the story.

Pasternak's remark and the limitations on a work's information content and complexity bring to mind Gödel's famous first theorem about the incompleteness of formal mathematical systems: In any sufficiently rich formal system there will always be statements that will be neither provable nor disprovable. Chaitin has shown that Gödel's theorem follows from the fact that no program can generate a sequence with greater complexity than it itself possesses. (Like the convoluted fractals generated from short and simple equations, however, a sequence may appear to be much more complex than the program that generates it.) As Chaitin has remarked,* we can't prove a ten-pound theorem (generate a very complex sequence) with five pounds of axioms (with a less complex program), and this limitation afflicts any information-bearing entity, human, electronic, or other.

The conception of scientific theories as programs generating sequences has many other limitations; it is quite simplistic and only begins to make sense relative to a well-defined and already fixed scientific framework. To apply these ideas to real science or to literature we would need a level of analysis the units of which were not bits but

---

*Germane to this observation is the so-called Berry paradox, which directs: "Find the smallest whole number which requires in order to be specified more words than there are in this sentence." Examples such as the number of hairs on my head, the number of different states of the Rubik cube, and the speed of light in millimeters per millennium, each specify, using fewer than the twenty words in the given sentence, some particular whole number. The paradox becomes apparent when we note that the sentence does specify a particular whole number which, by definition, it contains too few words to specify. Although Chaitin's proof of Gödel's theorem plays off of the Berry paradox, the incompleteness theorem itself is not in the least a paradox. It is strange, but it is real and unproblematic mathematics.

higher-order regularities, different for the different realms.

Murray Gell-Mann has suggested in *The Quark and the Jaguar* and elsewhere that we adopt a definition of *effective complexity* that accords better with our intuitions about meaning and information content. He notes that what we normally prize is not the shortest program (or theory or analysis) generating some sequence (or observation or entity), but rather the shortest program generating the sequence's "regularities." He thus defines the effective complexity of a sequence to be the shortest program generating its regularities. If the entity of interest were a story rather than a sequence, presumably the regularities would derive from stringing together basic actions and plot elements into coherent narratives.

Gell-Mann's modified definition of complexity results in random sequences of bits or other basic elements—which, recall, have no regularities—being assigned an *effective complexity* of zero (despite their maximal complexity). Although the notion of regularities is problematic, I find this idea most welcome for extramathematical reasons. The high complexity of random sequences is not only counterintuitive but seems to imply the unattractive idea that a desideratum of life (greater and greater complexity) leads inevitably to death (maximal complexity, or randomness).

The switch to effective complexity also resonates well with our belief that sequences that evince both order and randomness have the highest meaningful content and hence the greatest effective complexity (despite their middling complexity). After all, an informal characterization of a story, for example, as being as informative as possible doesn't mean it is a random collection of words. (Sequences with low complexity, it should be added for completeness, also have low effective complexity.)

• • •

The complexity of our brain and DNA connects these ideas to this book's concern with the notion of self. If we think of DNA as something like a computer program directing the building of an embryo, then rough estimates of the complexity of the embryo program (which I'll pass over) reveal it to be grossly insufficient to delineate the trillions of connections in a human brain. These connections come largely from the experiences of a specific time and culture, and thus a large part of our identity is supplied to us by events outside our skulls. The intricate particularities of a brain's wiring have too much complexity to be the result of the DNA-driven program, which determines only the rough structure of the brain and its general patterns of response to the environment.

Another, more far-reaching conclusion can be drawn. However information is encoded in the brain, the brain's complexity—its factual knowledge, associations, reasoning ability—is necessarily limited. Once again, a rough number—three billion bits has been proposed—can be attached to it, but here the existence of the number is more important than its value. The reason is that any phenomenon in nature more complex than the human brain is by definition too complex for us to comprehend. Alternatively, we cannot make predictions (generate binary sequences) of greater complexity than (the information encoded within) our brains. Regularities may exist that provide a key to understanding the universe, but they may be beyond what I term the human brain's "complexity horizon" (a notion that with time will likely gain much greater currency).

In other words, there may be a relatively short "secret to the universe" program, a theory of everything having complexity, say, ten billion bits, that we're just too limited (i.e., too stupid) to understand. Although they differ ineradicably, both traditional religious and scientific approaches to a hoped-for theory of everything share the perhaps naive assumptions that such a theory can be found and that its complexity will be sufficiently limited to be understood by us. Why assume that?

An inchoate sense of these general cerebral limitations has always haunted me. When I was a kid I had the recurring fear that some great new scientific discovery or philosophical insight would be announced, and I would find myself "seven brain cells short" of understanding it. It would be just beyond my personal complexity horizon. As a result of that irrational terror and of having read that alcohol kills brain cells, I resolved to be a lifelong teetotaler. My understanding of brains and conceptual breakthroughs has grown a little more sophisticated over the years, but the behavior has persisted.

(I should interject here that in addition to the limited complexity of our brains, there is another reason that regularities may be beyond our capacities; this time the constraint is physical rather than information-theoretic. John Horgan in *The End of Science* writes about the necessarily speculative nature of much theory making in modern science, especially in physics. The energies required to test theories are too enormous and the distances and masses too infinitesimal to yield experimentally verifiable results. He calls the science which results "ironic science" and compares it to art, philosophy, or literary criticism in offering nothing more than possible, interesting new ways to view the world. Is our universe one of many? Do quarks have parts? What is the real meaning of quantum mechanics? Such questions cannot be answered empirically, and lead, according to Horgan, to various just-so stories and idle conjectures.)

Finally, since any comprehensible entity is, by complexity theory, of less complexity than we are, such an entity is not, for that very reason, an appropriate deitylike entity. People generally don't worship that which is simpler than they are. This natural reluctance to deify the simple (except possibly as a symbol) is consistent with a tendency among some to identify God with the great unfathomable, the incomprehensibly complex. Reversing the letters in God and the lines of this thought, we note that it is also consistent with a dog's deifying its master (assuming, that is, that the master is of greater complexity than the dog).

Always a Free Lunch if the (Ramsey) Dump Is Big Enough

If we define God to be the incomprehensibly complex, then even agnostics and atheists can aver that they believe in such a God. This verbal trick does have a certain appeal, the appeal of getting something—God, in this case—for nothing. In this sense the subterfuge is related to the idea of order (or should it be Order) for free.

I've always been captivated by the notion that whatever the chaos in the collage of life, there will inevitably be pattern or order of some sort on some level. Since a *lack* of order or pattern is also a kind of (higher-level) order or pattern, the assertion of the inevitability of order is empty, vacuous, tautologous, but it is, I think, a very fruitful tautology. It's impossible that we be unable to point to some regularity, some invariant somewhere, no matter what the jumbled details of any particular state of affairs. This a priori disinterested view of the world that mathematics seems to foster is part of what drew me to the subject.

The idea of the inevitability of order is not without resonance in literature. In *Alice in Wonderland,* for example, after reciting a hilariously mangled version of the poem "Father William," Alice seems to intuit that a totally disordered world is logically impossible.

> "That is not said right," said the Caterpillar.
>
> "Not *quite* right, I'm afraid," said Alice timidly; "some of the words have got altered."
>
> "It is wrong from beginning to end," said the Caterpillar decidedly, and there was silence for some minutes.

Also apt is travel writer Pico Iyer's penetrating comment on the city of Bombay: "Everything goes wrong, and everything's all right."

In physics the idea of the inevitability of order arises in the kinetic theory of gases. There, an assumption of dis-

order on one formal level of analysis—the random movement of gas molecules—leads to a kind of order on a
higher level—the relations among macroscopic variables
such as temperature, pressure, and volume known as the
gas laws; the latter lawlike relations follow from the lower-
level randomness and a few other minimal assumptions.
More generally, any state of affairs, no matter how disorderly, can simply be described as random, and ipso facto,
at a higher level of analysis we have at least one useful
"meta-law": There is randomness on the lower level.

In addition to the various Laws of Large Numbers studied in statistics, a notion that manifests a different aspect
of this idea is statistician Persi Diaconis's remark that if
you look at a big enough population long enough, then
"almost any damn thing will happen." Also insufficiently
appreciated, even by some social scientists who should
know better, is the fact that if one searches for a statistical
correlation between almost any two random quantities in
a population, one will find some statistically significant association. It doesn't matter if the quantities are religious
affiliation and neck circumference, or (some measure
of) a sense of humor and job status, or perhaps the
amount of sweet corn consumed annually and the number of years of school completed. Moreover, despite the
correlation's statistical significance (i.e., its unlikelihood
of occurring by chance), the correlation is not likely to be
practically significant owing to the presence of so many
confounding variables. Nor does the correlation necessarily validate the often ad hoc story that accompanies it
that purports to explain why people who eat a lot of corn
go further in school. Superficially plausible tales are al-

ways available: corn eaters are more likely to be from the upper Midwest where dropout rates are low.

A more profound version of this line of thought can be traced to British mathematician Frank Ramsey, who proved a theorem stating that for a sufficiently large set of elements (people or numbers or geometric points) every pair of whose members are, let's say, either connected or unconnected, there will always be a large subset of the original set with a special property. Either all of the subset's members will be connected to each other, or all of its members will be unconnected to each other. This subset is an inevitable island of order (there are more interesting islands in the archipelago) in the larger unordered set; it is the free lunch guaranteed to exist if the garbage dump is large enough.

The problem may be phrased in terms of guests at a dinner party. The Ramsey question for the island of order of size 3 is: What is the smallest number of guests who need to be present so that it will be certain that at least 3 will know each other or that at least 3 will be strangers to each other? (Assume that if Martha knows George, then George knows Martha.) That the answer is 6 may be seen by pretending that you're a guest at the party. Since you know or don't know each of the other 5, you will either know at least 3 of them or not know at least 3 of them. Why? Suppose that you know 3 of them (the argument is the same if you don't know at least 3) and consider what relationships hold among your 3 acquaintances. If any 2 know each other, then these 2 and you constitute a group of 3 guests who know each other. On the other hand, if none of your 3 acquaintances know each other, they constitute a group of 3 who don't know each other. Thus 6 guests are sufficient. To see that 5 guests are not enough, pretend you're at a smaller party where you know exactly 2 of the other 4 guests, and each knows a different 1 of the 2 people you don't know.

For the island of order of size 4, the number of guests necessary is 18, and for 5, between 43 and 55. For larger numbers the analysis gets much more complicated, and answers to Ramsey-type questions are known for very few numbers.

Since Ramsey died in 1930, a whole cottage industry has grown up devoted to proving theorems of the same general form: How big does a set have to be so that there will always be some subset of a given size that will possess some regular pattern, an island of order of some sort? The prolific and peripatetic mathematician Paul Erdos discovered many such islands, some of them ethereally beautiful. The details of the particular islands are complicated, but in general the answer to the question about the necessary size of the set often boils down to Diaconis's dictum: if it's big enough, almost any damn thing will happen. As I suggested in the segment on biblical codes, Ramsey-type theorems may even be part of the explanation for some of the equidistant letter sequence results. Any sufficiently long sequence of symbols, especially one written in the restricted vocabulary of ancient Hebrew, is going to contain subsequences that appear meaningful.

In biology Stuart Kauffman has proposed in *At Home in the Universe: The Search for Laws of Self-Organization and Complexity* the related notion of "order for free." Motivated by the idea of hundreds of genes in a genome turn-

ing on and off other genes and the order and pattern
that nevertheless exist, he asks us to consider a large collection of 10,000 lightbulbs, each bulb having inputs
from two other bulbs in the collection. Subject only to
this constraint, one connects these bulbs at random. Assume also that a clock ticks off one-second intervals, and
at each tick each bulb either goes on or off according to
some arbitrarily selected rule. For some bulbs, the rule
might be to go off at any instant unless both inputs are on
the previous instant. For others it might be to go on at
any instant if at least one of the inputs is off the previous
instant. Given the random connections and random as-
signment of rules, it would be natural to expect the collection of bulbs to flicker chaotically with no apparent
pattern.

What happens, however, is that one observes "order for
free"—more or less stable cycles of light configurations,
different ones for different initial conditions. As far as I
know, the result is only empirical, but I suspect it may be
a consequence of a Ramsey-type theorem too difficult to
prove. Kauffman proposes that some phenomenon of
this sort supplements or accentuates the effects of natural
selection. Although there certainly is no need for yet another argument against the seemingly ineradicable silli-
ness of "creation science," the lightbulb experiments and
unexpected order that occurs so naturally in them do
seem to provide one.

A variant of the idea of order for free arises in philosophy in the pragmatic justification of induction. I mentioned in the appendix to chapter 3 what usually is
referred to as David Hume's traditional problem of in-

duction. Every day of our lives we confidently use inductive arguments (arguments the conclusions of which go
beyond, contain more information than, the premises).
Why, Hume asked, are we so confident that these arguments will yield true conclusions from true premises most
of the time? There certainly is no deductive argument
that since the sun has risen regularly in the past it will
probably rise tomorrow, or that since stones, when
dropped, have always fallen in the past they will probably
do so in the future.

It seems that the only argument for the continuation of
these sorts of regularities is an inductive one: since these
regularities have obtained in the past, they will probably
continue into the future. Yet attempting to justify the use
of inductive arguments by inductive argument clearly
begs the question. To put the matter baldly, the answer to
the question, Why will the future be like the past in certain relevant respects? seems to be nothing more satisfying than: It will be because past futures have been like
past pasts in these respects. This is helpful, though, only
if the future will be like the past, which is the point at issue. A Hume-ongous problem indeed!

Many attempts have been made to clean up this so-
called scandal of philosophy. One way out is simply to accept a nonempirical principle of the uniformity of nature
over time. The problem with this "solution" is that it too
begs the question: it is equivalent to what is to be estab-
lished; it has the advantage, as Russell said, of "theft over
honest toil." Another attempted way out is to note that
some inductive arguments are of higher level than others
and to try to make use of this hierarchy (of inductive ar-

guments, meta-inductive arguments, meta-meta-inductive arguments, and so on) to somehow justify induction. This doesn't quite work, or rather it works too well, and "justifies" a lot of weird practices, including counterinductivism. Some have even tried to dissolve the problem by showing that following our commonsense inductive rules is what is meant by rationality and no further justification is called for.

Charles Saunders Peirce and Hans Reichenbach advanced a different pragmatic justification of induction. It amounts roughly to this: Maybe induction does not work, but if anything does, induction will. Maybe there is no persisting order in the universe, but if there is any on any level of abstraction, induction will eventually find it (or its absence) on the next higher level. Although there is a problem with the word *eventually*, there is some merit and appeal to such an approach, which seems, as I've mentioned, different from but compatible with the idea of the inevitability of order.

Sᴛᴏʀɪᴇs, Aɴᴀʟᴏɢɪᴇs, ᴀɴᴅ Oʀᴅᴇʀ ꜰᴏʀ Fʀᴇᴇ

Since a theme of this book has been forging analogical beams to aid in the bridging of cultural gaps, we might ask, What is the relevance of all this to stories? Surely there are elements of order or recurring pattern that come almost for free here as well. A literary theorist is not required, for example, to account for the natural prevalence of the schematic love story: Boy meets girl, obstacles intervene, boy and girl get together. Sex, the disruptive nature of the world, and normal perseverance are sufficient to explain it. Likewise with the natural prevalence of stories of birth, journeys, and death.

Are there, in addition, closer analogues in literature and the human sciences of the order-for-free insights of the abovementioned disciplines? There most certainly is a natural correlate to the gas laws of physics, and as with many technical notions, it predates them considerably. More than 2,000 years ago the Roman poet and philosopher Lucretius wrote the following.

> For in very truth, not by design did the first beginnings of things place themselves each in their order with foreseeing mind, nor indeed did they compact what movements each should start, but because many of them shifting in many ways throughout the world are hurried and buffeted by blows from limitless time, by trying movements and unions of every kind, at last they fall into such dispositions as those whereby our world of things is created and holds together.

Another natural analogue of gas laws in physics is a more numerical version of Lucretius's basic idea. The nineteenth-century Belgian scholar Adolphe Quetelet argued that probability and statistical models can be used to describe social, economic, and biological phenomena; out of our unguided comings and goings emerge a certain pattern and frequency of crime (among other regularities). He wrote:

> Thus we pass from one year to another with the sad perspective of seeing the same crimes reproduced in the same

order and calling down the same punishments in the same proportions. . . . We might enumerate in advance how many individuals will stain their hands in the blood of their fellows, how many will be forgers, how many will be poisoners, almost as much as we can enumerate in advance the births and deaths that should occur.

Since Quetelet, we've been inundated with a torrent of statistical analyses of our births, deaths, health, incomes, and spending.

Nor need we go far afield to find evidence for Diaconis's pronouncement. It is borne out on the evening news and afternoon talk shows where, utterly unamazingly, amazing stories are reported almost every day. And one human illustration of Ramsey theory is that given a large enough population, we are guaranteed the existence of collections of people with unusual sets of relations holding among them.

I think it is remarkable that esoteric mathematical research such as Ramsey theory or Chaitin's work increasingly has some echo in the "real world." One more such echo deriving from articles on Ramsey-type theorems involves so-called phase transitions. The gist of these papers is that some combinatorial phenomena rarely occur until a certain critical number is reached, after which time they rarely fail to occur. Whether certain contemporary dysfunctions have arisen because the requisite critical number of connections among people (via mass media) have formed is intriguing. As seems to be true for genes, this interconnectedness may be critical. At a sufficiently high level of abstraction, such interconnectedness may give rise to all sorts of fancy periodic regularities that are implicated in our "social evolution" or in the recurrence of fads and fashions, or even in the seemingly inexplicable suddenness of traffic jams. Needless to say, more research is needed to elucidate and establish such regularities.

Does the pragmatic justification of induction have a social parallel? Such a justification lies, I think, in our modern refusal in social theorizing, historical writing, and especially literature to postulate any sort of grand order. We're content to pick up small bits of pattern if ever and wherever they might be found. Samuel Beckett's writings, which have always struck me as vaguely mathematical, come to mind, as does Hugh Kenner's translation of a passage from Beckett's *Watt* into the computer language Pascal. The minimalist stories of Raymond Carver or Ann Beattie also seem somehow informed by this view of induction, as do many meta-fictional novels and short stories. To use Isaiah Berlin's terms (via Archilochus), we're more likely to be foxes who know many little things than hedgehogs who know one big thing. Indeed, this book's piecemeal, episodic structure is due to a similar impatience with lofty claims and simplistic theories.

Imposing and overly cohesive literary works such as a novel by Thackeray, for instance, seem naive by contrast. As suggested earlier, the stream-of-consciousness novels of James Joyce and Virginia Woolf in the early part of twentieth century can be seen as the beginning of an attempt to discern pattern on one level by simply describing the most mundane actions and thoughts on a lower level. The story, such as it is, develops naturally out of scraps of experience and haphazard hitherings and thitherings. Narrative analogues of temperature (fraught with emotion), pressure (densely detailed), and volume (extensive) might even be sought if one has a high enough tolerance for sometimes tenuous analogies.

•   •   •

Another correspondence between stories and information is suggested by the notion of "physical entropy," defined by physicist Wojciech Zurek in his "Thermodynamic Cost of Computation, Algorithmic Complexity, and the Information Metric," which appeared in the September 1989 issue of *Nature.* Zurek defined physical entropy to be the sum of Claude Shannon's information content, measuring the improbability or surprise inherent in a yet to be fully revealed entity, and Chaitin's complexity, measuring the algorithmic information content of what's already been revealed. As science writer George Johnson shows in *Fire in the Mind,* this definition was designed to clarify certain classical problems in thermodynamics (in particular that of Maxwell's demon); but physical entropy can also be used to model the human-story system. Imagine two readers encountering a new short story or novel. One is a very sophisticated litterateur, while the other is quite naive. For the first reader the story holds few surprises, its twists and tropes being old hat. The second reader, however, is amazed at the plot, the characters, the verbal artistry. The question is, How much information is in the story?

In attempting to answer this, it makes sense to consider the readers as well, readers who bring vastly different backgrounds to the story. The first reader's mind already has encoded within it large portions of the story's complexity; the second reader's mind is relatively unencumbered by such complexity. The Shannon information content of the story—its improbability or the surprise it engenders—is thus less for the first reader whose mind's complexity is, in this regard, greater, whereas the opposite holds for the second reader. As they read the story both readers' judgments of improbability or surprise dwindle, albeit at different rates, and their minds' complexity rises, again, differentially. The sum of these two— the physical entropy—remains constant and is a measure of the information content of the mind-story system.

There are three aspects of this admittedly vague speculation that I like. One is that the notions involved are in the same conceptual ballpark as the second law of thermodynamics, which C. P. Snow used to illustrate the gulf between scientific and literary elites, the latter presumed not to understand the significance of the second law. Since this book in part is also a more oblique look at the gap between these two cultures, the historical resonance is satisfying. (For those interested, the probabilistic definition of Shannon's information content is similar to that of thermodynamic entropy in the second law, although the viewpoint one takes in using it is quite different: thermodynamic entropy increases in a closed system, whereas the information content of a message decreases as it is decoded.)

More important, such a speculation also provides a sense (not incompatible with some neuroscientists' theories, for example, or even with Hamlet's metaphorical utterance "Within the book and volume of my brain") in which stories become a physical part of us. They become encoded somehow into fragments of mental routines capable of generating them at will, and, if integrated into our conceptual and emotional maps of the world, they change us forever. We are the stories we tell.

The third appealing aspect of this line of thought is that it seems to underline the extent to which context (in this case the human-story system rather than just the story) is needed when making assessments. A story makes no sense to people who don't have any of the relevant linguistic and psychosocial understandings it presumes. Without a scientific and cultural matrix that supports these essential but implicit understandings, theories and stories are meaningless.

Perhaps this is what postmodernist literary theorists mean when they refer to "the death of the author." Their reluctance to accept the author's view of his or her work as definitive (or in extreme formulations, as even very important) may indicate a realization (or an overestimation) of the fact that meaning is a socially mediated phenomenon made possible by common cultural understandings. Wittgenstein once remarked aptly, "That Newtonian mechanics can be used to describe the world tells us nothing about the world. But it does tell us something—that it can be used to describe the world in the way in which we do in fact use it." The same might be said about Jamesian novels or Seinfeldian situation comedies. None of this entails the death of the author, however. The author is more than a celebrity's "as told to" ghostwriter for the culture as a whole, although celebrities and cultures as a whole do tend to be rather inarticulate.

COMPLEXITY, CHAOS, AND SPARE CHANGE

I have a box of change in a drawer that is always threatening to overflow, so each morning I take a handful of coins and resolve to spend them during the day. I hate to be without pennies when the cashier asks for $2.61 and threatens to leave me with the dreaded four pennies of change. Sometimes when through a particularly deft purchase I manage to get rid of all my pocket change in one transaction, I find myself happier than if I had accomplished something substantive that day. I also find myself making assumptions about items' prices (the prevalence of items for 99 cents, for example), the number of coins in my pocket, their denominations, the applicability of the Central Limit Theorem and other mathematical arcana, and then deriving small theorems about how often my pockets will be devoid of change. I'll spare us the particulars.

More trivia. The deadline is tomorrow for an article I promised to deliver but now don't want to write or for household bills that need to be paid or for some other less than exhilarating chore. I dutifully, if reluctantly, begin to work on the project when a niggling detail about some utterly irrelevant matter comes to mind. It may concern the etymology of a word, or the colleague whose paper bag ripped open at a departmental meeting revealing a pornographic magazine inside, or why caller ID misidentified a friend's telephone number, but I spend the next half hour trying to clear up my confusion about the matter. The misidentified caller ID then reminds me of someone I know who thought she was using the privacy button on her phone and was chagrined to discover that her rude characterization of her caller was overheard by the caller. This brings to mind the e-mail replies sent to the wrong sender, the colleague's lame explanation that he must have picked up the wrong bag, the origin of *lame*, and so on. (I've always liked Laurence Sterne's eighteenth-century novel *The Life and Opinions of Tristram Shandy, Gentleman*, which concerns the life of the raconteur narrator, Tristram Shandy, whose digressions were so extensive that it took him two years to write about the first two days of his life. Often impatient, I also find it to be a most annoying book.)

These prosaic episodes strongly suggest that there never will be *precise* sciences of happiness, efficiency, or human behavior generally, notwithstanding the efforts of utilitarian philosophers, hedonic psychometricians, ergonomics experts, and cognitive scientists. Novels and garden-variety conversations always have the advantage whenever human particularities are at issue.

An obstacle to precision in the putative sciences just mentioned is the utter complexity of the associations and linkages in our brain and the consequent chaos to which these sometimes give rise. Regarding the latter, a technique devised by topologist Steve Smale to illustrate the evolution of mathematical chaos is relevant. Imagine a cubical piece of white clay with a thin layer of red dye running through its middle. Now stretch and squeeze this cube to twice its length, then fold it smoothly back on itself to reform the cube. The dye layer is now shaped like a horseshoe. Repeat this stretching, squeezing, and folding many times and you'll notice that the red dye soon is spread throughout the clay in a most convoluted pattern similar to that in fancy lace or a filigreed pastry. Points in the dye that were close are now distant; other points that were distant are now close. The same is true for points in the clay. It has been argued that all chaos (and the consequent skittish unpredictability, disproportionate responses, and so-called Butterfly effects it gives rise to) results from such stretching, squeezing, and folding in a suitable logical space.

As I've written elsewhere, reading magazines and newspapers, watching television, or simply daydreaming and free-associating are efficient means for doing to our minds what the stretching, squeezing, and folding does to the red dye. The stretching and squeezing correspond to our envisioning of distant events, disparate people, and unusual situations (such as the time-wasting digressive thoughts I mentioned earlier). The folding corresponds to what we do if we try to connect these events, people, and situations with those of our own lives. Every day our mental landscape is stretched, squeezed, and if we allow it, folded back on itself, and the effect on us is similar to that on the red dye. Ideas, associations, and beliefs that were close become distant and vice versa. People keenly attuned to the world and what they read and see are much harder to predict, I suspect, than those whose range and purview are more limited.

This metaphor needs to be developed if it is to make any scientific claim, and as stated, is almost unfalsifiable. Still, it is suggestive and seems to be consistent with the idea broached earlier that we ourselves are so-called non-linear dynamical systems subject at times to the same chaotic unpredictability as New England weather. It does seem, for example, that dark moods come over us in the way bad storms bring a sudden end to walks in the park.

Of course, we're not that unfathomable, so there must be countervailing statistical and other considerations that make for predictability and stability. And yet I don't feel quite so pathetic when I interrupt a project to check on some obscure web site or newsgroup or derive an iota of cheer by getting rid of a pocketful of change.

•   •   •

The payoff to these analogies lies in a broader and more suggestive set of referents, scientific as well as literary. Biculturalism ought to reign in our individual cognitive homes, and souvenirs of travels and travails in foreign disciplines are one way to help further this. I am aware that part of what is written here may be dismissed as an unholy mixture of discordant fields; even I think this on Tuesdays and Saturdays. Nevertheless, on the other five days I think it's well worth a scientist's effort to try to explore the borderland between these disparate cultures. The alternative is to surrender it completely to postmodernists, phenomenologists, race-gender theorists, poststructuralists, Marxists, historicists, and psychoanalytic deconstructionists. Such people are, at least on Tuesdays and Saturdays, not without their charm and insights; but as physicist Alan Sokal's hoax* in the literary periodical *Social Text* intimated, the other five days they are more likely to bring forth balderdash disguised as profundity.

---

*Sokal's nonsensical paper, full of high-sounding ideas from physics and seeming to lend prestigious support to various relativist arguments, was published in the abovementioned journal.

5
βridgiŋg the ɢap

We do not have too much intellect and too little soul, but too little intellect in matters of the soul.
—**Robert Musil**

An INTEGRATION OF STORIES AND STATISTICS, or more generally, of the literary and the scientific, can be energizing. The drama and humanity of stories enhance scientific and statistical studies, while the rigor and disinterested perspective of the latter keep stories from degenerating into maudlin trifles or pompous puffery. Metaphor and analogy stretch the narrow literalness of mathematical and scientific understanding, and mathematical calculations and constraints ground literary imagination.

The issue is *not* literary imagination versus scientific substance. Stories frequently are more pivotal to understanding not only ourselves but even mathematics and science than are formulas, equations, and statistics; and mathematical and scientific ideas frequently are more creative and visionary than novels or plays. Were I given to hyperbole I might refer to my forays across the story—

statistics divide as a mathematician's intellectual trans-
vestism. Since I am not so inclined, instead let me sound
a warning about the *indiscriminate* merging of the narrative and the numerical.

Such facile merging can sometimes signal a philosophical confusion, or what the British philosopher Gilbert
Ryle long ago termed a category mistake. One such error
tells of the peasant who asks the master whether *kebab* is
with an *a* or an *o*; the master responds that it's with meat.
Anecdotes provide a more relevant instance of skirting
the border between stories and statistics, and when used
to advance arguments instead of merely illustrating them,
often constitute an illegal crossing.

Another often misguided attempt at integrating the
two realms is to dress statistical and social-scientific analyses in novelistic garb. In Tom Wolfe's anthology *The New
Journalism*, for example, he discusses techniques that the
then new journalism had borrowed from the novel: using
dialogue to replace reported speech; telling the story
through dramatic scenes and scenarios rather than
through expository summaries and statistics; assuming a
particular attitude and point of view rather than a disinterested, impersonal perspective; and making details of
clothing and appearance critical rather than part of a
generic background. Other techniques involve verb tense
(frequently the historic present or simple past) to create
the illusion of eavesdropping, raising questions and de-
laying answers to create suspense, and the occasional use
of crosscuts and flashbacks for dramatic effect.

In short, the techniques make it easier to view reality as
just another story; they blur the boundary between novels

and nonfiction, and increasingly, between entertainment and news. The status of nonfiction accounts and the news frequently is not privileged. The small detail that, given due allowance for fallibility, vagueness, point of view, and so on, they report what actually happened, is often lost in the shuffle. Conversely, novelists should refuse to conceive of their work as propaganda for Marxism, say, or any other totalizing "ism" or even as social-scientific analysis. That they almost always do refuse is probably why novels are disdained by those social activists who, impatient with individual visions, recognize only grandiose collective movements.

An improper conflation of the personal and the impersonal also may arise when we try to project our life circumstances onto an indifferent world. Consider, for example, the inclination of old or sick people to elevate their personal stories of loss into apocalyptic admonitions of social decline; starkly phrased, this amounts to: "I'm decaying so the world must be too." It doesn't take much psychological acuity to realize that many millennialists and apocalypticians actually *want* the world to end. The world's coming to an end when you do is self-enhancing; it can be dispiriting to viscerally realize that life will go on without you.

Equally off-course is the tack taken by some chirpy technophiles who so identify with the impersonal progress of science as to be oblivious to their own individual stories and plight. Such opposing tendencies probably underlie the remark that the movement from the Enlightenment to Romanticism was one from objective optimism to subjective pessimism. Science progresses in a pleasant, impersonal way, whereas individuals inevitably decompose.

As these chapters attemp to show, the gaps separating stories and statistics, subjective viewpoint and impersonal probability, informal discourse and logic, and meaning and information are bridgeable in places, not so in others, and seldom well-marked. Nowhere are the gaps more indistinct than at the border between some of the disparate realms briefly touched on here.

LOTTERIES AND TURNING WISHES INTO FACTS

I've been interviewed on countless radio and television shows, and the conversation frequently turns to lotteries and ways to beat them. On one such show I was told that in some states hand-picked lottery numbers won more often than did machine-generated numbers. I was tempted to dismiss this as another instance of silly lottery lore, but I realized that although difficult to check, the claim is not necessarily nonsense. In fact, it nicely illustrates one way in which personal wishes can sometimes seem to affect large, impersonal phenomena.

Consider the following simplified lottery. In a comically small town, the mayor draws a number from a fishbowl every Saturday night. Balls numbered from 1 to 10 are in the bowl, and only two townspeople bet each week. George picks a number between 1 and 10 at random. Martha, on the other hand, always picks 9, her lucky number. Although George and Martha are equally likely to win, the number 9 will win more frequently than will any other number. The reason is that *two* conditions must be met for a number to win: it must be drawn on Saturday night by the mayor and it must be chosen by one of the participants. Since Martha always picks 9, the second condition in her case is always met, so whenever the mayor draws a 9 from the bowl, 9 wins. This is not the case with, say, 4. The mayor may draw a 4 from the bowl, but chances are George will not have picked a 4 so it will not often be a winning number. George and Martha have equal chances of winning, and each ball has an equal chance of being chosen by the mayor, but not all numbers have the same chance of winning.

Are there other, less structured situations that resemble this one in relevant ways? Astrological beliefs come to mind. If enough people believe in such celestial balderdash and begin to mold their behavior to conform to what they believe to be their "true" natures, then their beliefs might be mildly self-fulfilling (despite the vagueness of astrological profiles). Similar observations can be made about orthodox Freudianism, creation science, and other closed belief systems. But might we more closely approximate the numbers-in-a-bowl example? Are there natural "lotteries" in which chance draws the ball from the bowl and some people randomly pick different numbers for each drawing and others consistently pick the same numbers for whatever reasons?

Consider the outcome of complicated but quite discouraging medical situations to be the analogue of lottery results. Assume that with some regularity patients are critically ill with many systems failing, and the particular course of their conditions is impossible to predict. The possible outcomes might number 10 with only number 9 corresponding to a recovery, the others corresponding to different ways in which the patient might die. In this case Martha's always "betting" on the patient's recovery owing to her belief in prayer might correspond to her always picking her lucky number 9, whereas George's nonbelief in prayer and his guess of a particular pathway to death might correspond to his picking a number at random. Even though belief in prayer would be just as unwarranted as any other expectation in this situation, recoveries would be accurately predicted considerably more often than would, say, death by pulmonary embolism.

More generally, any situation having many outcomes, one of which is conventionally favored by the society at large, will seem to generate the conventionally favored outcome more frequently than chance would suggest. This in turn will tend to make that outcome even more of a conventional favorite. In this attenuated sense, wishes and beliefs do become facts.

●   ●   ●

The stock market is also a kind of lottery, but the large number of traders gives it a decidedly different, more jittery flavor. Its size and complexity offer even more scope for turning wishes into facts and justify taking at least two very different stances toward it. Do we invoke only statistics—mathematical notions such as random walks, efficient markets, and beta values—or are narrative ingredients—such as fear, exuberance, and disproportionate response—part of the picture? Again, some straddling of the story-statistics line is the most appealing option. The failure of most highly paid fund managers, with all their charts and figures, to do better than blind indexed funds is something of a scandal and suggests that the theoretical statistical arguments are sound. Given a level of risk, only a certain level of return can be expected over the long term. Why try to outsmart dice?

On the other hand, the normal bell-shaped distribution does not accurately reflect the extreme volatility often displayed by the market. Investor psychology and the lockstep behavior it sometimes engenders (compare the parable of the furious feminists and the idea of probabilistic coupling discussed earlier) also suggest that to a limited extent, the market can itself be considered a kind of agent. A better term than agent is complex adaptive system, an important but rather technical notion studied by Per Bak, Brian Arthur, and other researchers at the Santa Fe Institute. Complex adaptive systems—of which we ourselves are instances—eventually may help clarify the nature of the gap between stories and statistics. It might even turn out that we're not always being absurdly anthropomorphic when we tell stories about such systems and their moods.

PROMISES, QUESTIONS, AND IMPLICIT INTENTIONS

Like wishes and fears, questions and promises sometimes play a surprising role in the creation of facts. Consider, for example, the different sort of traffic between objective and subjective that occurs in the following adaptation of a story by puzzle-meister Raymond Smullyan.

A man asks a woman, "Will you promise to give me a photograph of yourself if I make a true statement and, conversely, not give me a photograph if I make a false statement?" Feeling this to be a flattering and benign request, she promises. The man then states, "You will neither give me a photograph of yourself nor will you sleep with me." Spelling out the man's trick, I note that she can't give him a photograph of herself, since if she were to do so his statement would be made false and so she would have broken her promise to give him a photo only if he made a true statement. Therefore she must not give him a photograph under any circumstances. But if she also refuses to sleep with him, his statement becomes true, and this would require her to give him a photo. The only way she can keep her promise is to sleep with him so that his statement becomes false. The woman's seemingly innocuous promise ensnares her.

Fortunately or unfortunately, I suspect that the class of people for whom this seduction technique would prove effective is probably rather small. Nevertheless, it might make an interesting premise for a *Star Trek* episode, or perhaps form part of a logician's dating manual.

A case in which any definitive response to a question insures the answer's objective falseness is provided by the following classic story. Imagine a computer, call it Delphi-Omni-Sci (I always liked DOS), into which has been programmed the most complete scientific knowledge possible, the initial conditions of all particles, and elaborate mathematical techniques and formulas. Imagine further that Delphi-Omni-Sci answers only Yes or No questions and that its output device is constructed in such a way that a Yes answer turns off an attached lightbulb and a No answer turns on the lightbulb. If one asks this impressive machine something about the external world, the machine will respond, let us assume, flawlessly. However, if one asks the machine if its lightbulb will be on after answering the question, Delphi-Omni-Sci is stumped and cannot answer either way. If it answers Yes, the lightbulb goes off, while if it answers No, the lightbulb goes on. This question, at least, is "undetermined" by the laws and axioms of its program (although an onlooking computer might be able to answer it).

Related to Delphi-Omni-Sci is the following phenomenon familiar to parents of small children in particular. In predicting what a person will decide it is often very important to keep this predictive "information" secret from the person deciding else it change his decision. The scare quotes around "information" indicate that this peculiar type of information loses its value and becomes obsolete if it is given to the person whose decisions are being predicted. The information, while it may be correct and true, is not universal. The onlooker and the deciding agent (compare the interrogator and Delphi-Omni-Sci) have complementary and irreconcilable viewpoints. As D. M. MacKay has written in "On the Logical Indeterminacy of a Free Choice" in *Mind* (1962), our choice is indeterminate until we make it and not—as it is for an onlooker—something to be observed or predicted.

(This logical indeterminacy attaches to our predictions of what we ourselves will decide. In many situations we can avoid it by considering just parts of ourselves. In picturing situations involving us we can objectify those parts

of ourselves so involved. Our account of the situation is
then necessarily incomplete, however, since a part of us—
the subject observer—is always doing the observing and
predicting, and that part is not being self-observed or self-
predicted.)

In any case, a confusion or conflation of subject and
object (as in the case of Delphi-Omni-Sci) always results
in undecidable, open questions. There the conflation re-
sulted merely in the undecidability of certain questions
involving the attached lightbulb. Usually much more is
undecidable.

•   •   •

Not only these logical arcana but even garden-variety
conversations and intentional explanations (those which
make reference to the intentions and rationale of the
agent) involve such subject–object blurs. This is because
they require of us enough empathy to understand the
rules and constraints, values, and beliefs of another person
whose responses and actions are in turn thereby affected.
The philosopher H. P. Grice has even defined "S's meaning something by X" as "S's intending the utterance of X to
produce some effect in a hearer by means of the hearer's
recognition of S's intention to produce that effect." In a
sense we're all, at times, lightbulbs on other people's computers, and the cognitive coupling that results requires, as
maintained earlier, a new intensional logic.

Consider the following little interchange.

GEORGE: This talk of motives, promises, questions, fears,
and wishes is sloppy. Why don't we simply cite facts, use

> extensional logic, do mathematics and forget all this messy stuff?
>
> MARTHA: I know what you mean. Let's just vow to each other to do that from now on. We both yearn for clarity and precision.

The joke (such as it is) is that George and Martha are planning to use only extensional logic, but intensional notions are built into the very fabric of their (and our) communication. The situation is rather like shouting about the importance of silence, or claiming that one's brother is an only child. Although wishes, fears, promises, and motives are not (yet) the province of mathematics and science, they and the stories in which they're embedded are an essential ground for understanding these subjects and their applications.

TWO CULTURES, SAME PAROCHIALISM

One reason for writing about the relationship of stories to statistics is that it mirrors the more general relationship between C. P. Snow's two cultures, the literary and the scientific, without arousing all the standard knee-jerk platitudes that discussions of his 1959 talk usually do. Unfortunately, the chasm between these two cultures persists, each continuing to hold the other in mild contempt. Many literary types publicly speak and write as if they are the only kind of intellectuals, while many scientific thinkers secretly believe that much of the scholarship in literature and the humanities is muddleheaded and pretentious blather.

Although practitioners on each side tend to be both elitist and parochial, literature is naturally engaging and has a history and tradition that most anyone can relate to. Mathematics and science, on the other hand, too often are presented as a bagful of arcane techniques that sprang full-blown out of no one's told where. The implicit pedagogical strategy in many mathematics and science classes remains: Just shut up and do the problems. (This is not to say that there should be no computational drills, but merely to remind so-called math fundamentalists that computational facility is a vastly overrated skill, especially nowadays. Just as no one would confuse a good speller with someone who writes well, no one should equate a whiz at arithmetic with someone who understands and can effectively employ mathematical ideas.)

As a result of the pedagogical obsession with technique, the conceptions students in these classes and people in general have about the nature of mathematics and science often are quite narrow (and would be even more so were it not for the flourishing state of popular science writing). My students, for example, commonly act as if they've been betrayed when I require of them a concise, well-written expository paper, and people I encounter socially often seem to think it a little peculiar that I write.

Inexcusably, mathematics and science courses often omit any mention of the history or culture of these fields, seldom provide an overview of their major ideas and applications, and rarely allow for any meta-discussion of the subjects. Literature and humanities courses, on the other hand, often do little else and frequently come across to students as wordy, vacuous, unsystematic, and open-ended. (See "Poetry for Physicists" by Sheila Tobias and Lynne Abel in *The American Journal of Physics*, September 1990.) Such differences cannot be completely effaced (just as the related distinction between analytic and synthetic statements cannot be), but they should be explained so that students realize the many sorts of understanding and how to attain them.

The joke about the mathematics professor who gave a test consisting of four problems is apt. The first three problems required proofs of theorems and the last one was a statement prefaced with the directions "prove or disprove." One student toiled for a while and then came up to the professor's desk and asked, "On that last problem, do you want me to prove it or disprove it?" The professor responded, "Whichever is the right thing to do." "Oh," replied the student, "I can do either one. I was just asking which one you preferred." The interchange, of course, would not be a joke if the subject were history or literature.

THE DOCTOR'S DILEMMA

John J. McCarthy has illuminated these issues with a parable entitled "The Doctor's Dilemma."* McCarthy, one of the earliest researchers in artificial intelligence and inventor of the computer language LISP, is clearly a member of the scientific culture, and he asks us to contrast his response to the following dilemma with responses from members of the literary culture.

---

*It's on his web page at *www-formal.stanford.edu/jmc/docdil.html*.

The premise—which, ironically, requires a greater suspension of disbelief than does most fiction—is that a miracle has occurred and a young doctor working in a hospital wakes up and discovers that he has the power to cure all disease in anyone under the age of 70 with the briefest touch of his skin. A devoted doctor, he wants to make maximal use of his gift, since he knows it is nontransferable, will die with him, and will not inhere in skin removed from him. What should he and others do with this power?

Before giving his answer McCarthy lists a number of what he calls "literary exercises in paranoia." These include: the doctor cures people effortlessly and arouses the jealousy of his colleagues, who have him jailed and eventually lobotomized. A church charges him with sacrilege for violating the belief that man's inevitable lot is disease and death, and puts him to death after much suspense and intrigue. Alternatively, the gift is deemed holy and he is so hemmed in by beautiful ritual that he can make little use of his powers. In another vignette the doctor becomes so exhausted from his ministrations that he dies after delivering a poignant speech about his limitations. Or he might work hard at first and then succumb to an increasing desire for power, money, and women. (Note that unprotected sex is, for him, safe sex.) Then again, he could be taken over by the government, which might either keep him to cure high officials or appoint a board to insure that his powers are equitably applied to people from all racial and ethnic groups. The possibility of massive overpopulation and plagues due to the loss of immunities is imagined and taken as an argument to limit the doctor's access to patients.

In all instances drama and plots abound, controversies are stirred, and little curing is accomplished. McCarthy describes other scenarios involving the CIA and foreign powers, terrorists, mad scientists, mafiosi, and distraught parents, in each case mocking the dramatic embroidering and conspiracy concocting that he maintains characterizes most literary types.

Finally, McCarthy provides what he describes as a moral solution which, although it holds less literary appeal than the preceding scenarios, would enable the doctor to cure everyone under the age of 70, provided they are diagnosed in time and the doctor is still living. He maintains that his solution (or one like it) is much more likely to be devised by a member of the scientific culture (although it requires little knowledge of science). What is needed are a few numbers and an ability to do arithmetic. Say approximately 100 million people under 70 die each year, or slightly more than three per second. A machine consisting of 10 conveyor belts could be built, each belt moving 20 people per second past the doctor for a brief touch. He needn't work more than a half hour per day. McCarthy sketches objections and refinements to the method—machines in different parts of the world, overpopulation risk (none at all if the birthrate were to fall just marginally), and other peripheral moral and technological issues—but the core of the solution is a little hardheaded calculating.

Whether McCarthy's technological solution to his Doctor's Dilemma is of less literary merit than his "literary exercises in paranoia" is unclear, and in the presuppositions of its dramatic presentation reminds me a bit of the

George–Martha exchange described earlier. The article does, however, effectively caricature representatives of the literary culture. As Susan Sontag has commented in *Illness as Metaphor*, it is not uncommon for people to associate various story schemas and personality types with diseases ranging from tuberculosis to polio to AIDS. These schemas and types evaporate once a cure is found. Still, in the spirit of fair play and bridge building, I include here a less calculating point taken from a different Doctor's Dilemma, George Bernard Shaw's play of the same name: "Use your health even to the point of wearing it out. That is what it is for. Spend all you have before you die; and do not outlive yourself."

• • •

Another unvalidated test of the difference between the literary and scientific mind-set is provided by initial reactions to the death of Princess Diana. Was it viewed as some sort of drama, soap opera, or morality play with villains, heroes, and subtle plot twists, or was it viewed as an accident with very mundane antecedents such as excessive speed, drunkenness, poor judgment, and dangerous road conditions? In my informal canvassing of people soon after the accident, those with a scientific bent generally mentioned the latter considerations (or that this aficionado of psychics was badly let down by her benighted charlatans); those with more literary concerns (with several extreme exceptions) railed against the paparazzi or held forth upon the ironies of her life. Whatever their true distribution among the cultural types, the two positions are uneasy companions, but not incompatible. Revealingly, the stereotypical literary position is the more comforting. The various bathetic stories told about the paparazzi, love life, and royal family helped divert attention from a more distressing reality: the simple bad luck of an accident coupled with the usual human feelings of despair and impotence in the face of any death.

THE ENVIRONMENT AND OTHER NO-MAN'S-LANDS

Many forms of writing take place in the gray area between narrative and numbers, between literature and science. The phrase "no-man's-land" to describe this border area seems particularly fitting, since seldom are real characters about. Most popular science, a good deal of science fiction, some economic writings, certain kinds of philosophical work, and even pornography are in this no man's land. Of course, these writings contain some narrative conventions and pseudo-agents, but usually the stories provide only a minimal binding and serve primarily as display racks for the real attractions: the explanations and elucidations in the case of popular science writing; the premise and its consequences and the high-tech thingamajigs in science fiction; the explications of monetary and fiscal principles in economics articles; the philosophical points that parables and thought experiments are supposed to elucidate; and the sexual acts that are the staple of pornography.

A sort of Doctor's Dilemma writ large, writings on the environment sometimes occupy this gray area as well, and almost always do if they focus on the distant future. Sci-

ence, science fiction, economics, philosophy, and grue-
some, quasi-pornographic scenarios all play a role in fu-
turistic environmental projections in which real people
are necessarily replaced with cardboard cutouts. These
projections, like Rorschach tests, invite us to impose our
preconceptions on an amorphous, complex, and indifferent landscape. It is instructive to consider a couple of examples in which no greedy corporate despoilers or
zealous tree huggers play a part, and in which many but
not all of the elements of narrative are removed.

Assume first that on the eve of the millennium society
must make a major environmental policy decision and that
a determination to go forward with the policy proposal car-
ries much future risk. If adopted, initially there will be
some social disruption—people changing residence, much
building and construction, new organizations formed—
but the risky policy will lead to a significant increase in living standards for at least 300 years.

At some indeterminate time after that, however, a major catastrophe will occur, directly attributable to the
adoption of this risky policy, in which 50 million people
will die. (Imagine that perhaps the decision concerns the
disposal of radioactive wastes or building on a geologi-
cally unstable site.) Now, as English philosopher Derek
Parfit pointed out in *Reasons and Persons*, it could be
maintained that the decision to follow the risky policy was
bad for no one. The policy certainly wasn't bad for those
whose standard of living was increased for the centuries
before the catastrophe.

Moreover, the policy wasn't bad for the people who
died in the catastrophe, since they wouldn't have been

born were it not for the decision to follow the risky policy. This policy, remember, led to some initial upheaval and consequent altering of when existing couples conceived their babies (and hence of their babies' identities) and also, since different people were brought together, of which pairs become couples and then parents (and hence of their babies' identities). Over the course of centuries these differences multiplied and exfoliated and it could reasonably be assumed that no one alive the day of the catastrophe would have existed had not the risky policy decision been taken. The people who die, to reiterate, will owe their lives to the decision.

We thus have an example of a decision that leads directly to the death of 50 million people and yet is arguably bad for no one. What is apparently needed is some impersonal moral principle(s) in whose light we could reject the risky policy. Without the immediacy imparted by a story—a point-of-view and a narrator and nuanced actors with whom we can identify—most people would find it hard to care about the inevitable disaster. If a catastrophe happens in the distant future and no one hears it, . . . ?

Another example with a payoff 300 years into the future is provided by Steven Landsburg, author of *The Armchair Economist: Economics and Everyday Life*, who has argued that our descendants, despite our supposed plundering of the planet, will live incomparably better than we do and that our concern with the environment is sometimes excessive. Landsburg asks us to imagine a family of four earning the present median income of approximately $32,000. If, as economic history suggests is quite feasible, the U.S. per capita income were to grow *in real terms* at a quite reason-able real rate of 2 percent annually, then in just 300 years this family of four (well, not quite this one) will have an annual income over 12 million dollars! And these aren't inflation-shrunk dollars; they provide the equivalent of a 1997 income over 12 million dollars. With a higher rate of real growth, the time required to reach such an income could be much shortened.

The punch line is that each time an environmental group forestalls economic development, it is, in effect, asking contemporary ordinary working people who make a comparatively minuscule amount to sacrifice for the pleasure of future generations of multimillionaires. (And they will likely be longer-lived millionaires at that; in 1920 the average American life span was 54; in 1985 it had risen to 75.) This is the reverse of a progressive tax sys-tem, which allows the tax collector to take, let's say, 40 percent of large incomes. The spirit of such a progressive system, it would seem, would likewise allow unemployed lumberjacks to take part of our fabulously wealthy de-scendants' unobstructed view of the pristine forest. More-over, our descendants might well prefer the considerable proceeds of economic development of the area to the view of the forest. There are, needless to say, counterargu-ments.

Again, the abstractness of the issues, both scientific and moral, and the difficulty of imposing a conventional story on them tends not to attract those who prefer accounts of pitched battles between the forces of greed and good. In considering possible distant environmental futures, we seem to be cast adrift with only these abstract stories to guide us. These stories and standard conundrums such as the Prisoner's Dilemma and the Tragedy of the Commons are much better than nothing, but it's still a no-man's-land out there and we seem to want to foist an inappropriate narrative structure onto it.

A Word on (Ur)religion

Nirvana, the Garden of Eden, Shangri-la, Heaven. Narratives postulating the existence of such beatific "environmental" futures play a pivotal role in many religions. The creation myths, holy chronicles, and apocalyptic predictions of various religions certainly also count as vital stories. In the Bible, for example, the movement from Genesis to Revelations manifests characters, points of view, desires and fears, particularities of context, uniqueness claims, a direction in time, subplots, and every conceivable story element. Comparably majestic and full-blooded sagas appear in the Koran and the Bhagavad Gita. Despite their invaluable virtues, however, mathematics, statistics, and science generally lack life in the same sense—characters, plots or subplots, contexts, emotions, et cetera.

Religion can be thought of in part as an attempt to reconcile the personal with the impersonal by diminishing, if not eliminating, the latter. (Buddhism is an exception.) Physical processes, impersonal forces, and unlikely events are transmuted into personal actions, omniscient agents, and dark portents. Everything, including the potentially graspable and assumed-to-exist "meaning of it all," is understood to be part of a dramatic story.

(Finding it impossible to believe such stories, I've always wondered about the possibility of an ur-religion acceptable to atheists and agnostics. By this I mean a "religion" that has no dogma or narratives whatsoever and yet still captures something of the essential awe and wonder of things and affords as well an iota of serenity. The best I've been able to come up with is the "Yeah" religion, whose response to the intricacy, beauty, and mystery of the world is a simple affirmation and acceptance, "Yeah," and whose only prayer is the one word, *Yeah.* This minimalist "Yeah" religion is consistent with more complex religions [one exception being the "Nah" religion] and with a nonreligious ethics and a liberating, self-mediated stance toward life and its stories. Furthermore, it conforms nicely with a scientific perspective and with the idea that the certainty of uncertainty is the only kind of certainty we can expect.)

The physical sciences display a contrasting attitude toward stories. Many of their claims can be seen as an attempt to diminish, if not eliminate, the personal. One's most private feelings and attitudes, successes and failures, are described as no more than the consequences of some psycho-socio-econo-bio-chemo-physical generalization, and even the self and its feeling of I-ness can be explained as a kind of comical delusion, growing out of organisms' biological needs and the sheer unfathomability of their brains.

Ultimately, the gap between stories and statistics, and perhaps that between religion and science, may be a facet of the mind–body problem, that is, of the relationship between consciousness and physical stuff, a conundrum

whose various solutions, dissolutions, and stubborn reap-
pearances I won't even attempt to list. Whatever our religious feelings (or lack thereof) or scientific under-standing,
religious stories and scientific/statistical accounts can coexist in their disparate realms provided that simple-minded,
universal, and destructive attempts to reduce one to the
other are resisted. Some gaps, if they have any bridges
across them at all, should have only individual catwalks.

If and when one can get beyond the need for evidence
and empirical tests, the world's complexity provides am-
ple space for a variety of religions. There is more than
enough imaginative room beyond our collective
complexity horizon for all sorts of creation myths, quasi-
historical construals, afterlife narratives, and the traditions and ethical sensibilities that grow around them. For
some, a sort of intellectual compartmentalization may be
necessary to accept both science and religion, but to a
lesser extent this double vision is also needed in everyday
life, where we all juggle first-person and third-person
views of ourselves.

How we can maintain a place for the individual, pro-
tected from the overweening claims of religion, society,
and even science, is an increasingly important unsolved
problem. Its solution, I have no doubt, will require simply
and pragmatically accepting the indispensability of both
stories and statistics and of their nexus, the individual
who uses and is shaped by both. The gap between stories
and statistics must be filled somehow by us.

selected
Bibliography

Lists—of grocery items, things to do, appointments, etcetera—lie somewhere along the great continuum between narratives and numbers. Concerned with a somewhat nebulous complex of topics, the following bibliographic list is more than a bit arbitrary. The books below elaborate upon the more technical issues raised herein.

Applebaum, David. *Probability and Information.* New York: Cambridge University Press, 1996.

Barrow, John D. *Impossibility.* New York: Oxford University Press, 1998.

Barwise, Jon. *The Situation in Logic.* Stanford, Calif.: Stanford University Press, 1989.

Beckett, Samuel. *Watt.* New York: Grove Press, 1970.

Borges, Jorge Luis. *Labyrinths.* New York: New Directions, 1988.

De Botton, Alain. *How Proust Can Save Your Life.* New York: Pantheon, 1997.

Casti, John L. *Searching for Certainty.* New York: Morrow, 1990.

Chaitin, Gregory. *The Limits of Mathematics.* Singapore: Springer-Verlag, 1997.

Cuzzort, R.P., and James S. Vrettos. *Statistical Reason.* New York: St. Martin's Press, 1996.

Dennett, Daniel, C. *Darwin's Dangerous Idea.* New York: Simon & Schuster, 1995.

Devlin, Keith. *Goodbye, Descartes.* New York: Wiley, 1997.

Drosnin, Michael. *The Bible Code*. New York: Simon & Schuster, 1997.

Empson, William. *Seven Types of Ambiguity*. New York: New Directions, 1947.Gardner, Martin. *The Whys of a Philosophical Scrivener*. New York: Quill, 1983.

Gell-Mann, Murray. *The Quark and the Jaguar*. New York: Freeman, 1994.

Gould, Stephen Jay. *Full House*. New York: Harmony Books, 1996.

Haack, Susan. *Philosophy of Logics*. Cambridge University Press, 1978.

Hofstadter, Douglas. *Le Ton Beau de Marot*. New York: Basic Books, 1997.

Horgan, John. *The End of Science*. New York: Addison-Wesley, 1996.

Johnson, George. *Fire in the Mind*. New York: Knopf, 1995.

Kadane, Joseph B., and David A. Schum. *A Probabilistic Analysis of the Sacco and Vanzetti Evidence*. New York: Wiley, s1996.

Kauffman, Stuart. *At Home in the Universe*. New York: Oxford University Press, 1995.

Moore, David, and George McCabe. *Introduction to the Practice of Statistics*. New York: Freeman, 1993.

Parfit, Derek. *Reasons and Persons*. Oxford, U.K.: Clarendon Press, 1984.

Paulos, John Allen. His previous books are of some relevance to the matters discussed in this one.

Quine, Willard Van Orman. *Methods of Logic*. Holt, Rinehart, and Winston, 1959.

Ronen, Ruth. *Possible Worlds in Literary Theory*. Cambridge University Press, 1994.

Ross, Sheldon. *First Course in Probability*. New York: Macmillan, 1994.

Ruelle, David. *Chance and Chaos*. Princeton, N.J.: Princeton University Press, 1991.

Sterne, Laurence. *The Life and Opinions of Tristram Shandy, Gentleman.*

Sutherland, Stuart. *Irrationality: The Enemy Within.* Constable, 1992.

Turner, Mark. *The Literary Mind.* New York: Oxford University Press, 1995.

Tversky, Amos, and Daniel Kahneman. *Judgement Under Uncertainty: Heuristics and Biases.* Cambridge University Press, 1982.

ıŋdɛx